Jingzhu Zhao

Ecological and Environmental Science & Technology in China: A Roadmap to 2050

Chinese Academy of Sciences

Jingzhu Zhao
Editor

Ecological and Environmental Science & Technology in China: A Roadmap to 2050

With 16 figures

Science Press
Beijing

Springer

Editor
Jingzhu Zhao
Institute of Urban Environment, CAS
361021, Xiamen, China
Email: jzhao@rcees.ac.cn

ISBN 978-7-03-026651-4
Science Press Beijing

ISBN 978-3-642-12714-4 ISBN 978-3-642-12715-1 (eBook)
Springer Heidelberg Dordrecht London New York

Library of Congress Control Number: 2010925331

Cover design: Frido Steinen-Broo, EStudio Calamar, Spain

Printed on acid-free paper

Springer is a part of Springer Science+Business Media (www.springer.com)

Members of the Editorial Committee and the Editorial Office

Editor-in-Chief

Yongxiang Lu

Editorial Committee

Yongxiang Lu Chunli Bai Erwei Shi Xin Fang Zhigang Li Xiaoye Cao Jiaofeng Pan

Research Group on Ecological and Environmental Strategies of Chinese Academy of Sciences

Head: Jingzhu Zhao	Institute of Urban Environment, CAS
Members: Jingzhu Zhao	Institute of Urban Environment, CAS
Yongguan Zhu	Research Center for Eco-Environmental Sciences, CAS
	Institute of Urban Environment, CAS
Jiuhui Qu	Research Center for Eco-Environmental Sciences, CAS
Guobin Liu	Institute of Soil and Water Conservation, CAS
Guirui Yu	Institute of Geographical Sciences and Natural Resources Research, CAS
Huijun Wang	Institute of Atmospheric Physics, CAS
Yunfa Chen	Institute of Process Engineering of CAS
Guishan Yang	Nanjing Institute of Geography and Limnology, CAS
Xingguo Han	Shenyang Institute of Applied Ecology, CAS
Shenghui Cui	Institute of Urban Environment, CAS

Foreword to the Roadmaps 2050*

China's modernization is viewed as a transformative revolution in the human history of modernization. As such, the Chinese Academy of Sciences (CAS) decided to give higher priority to the research on the science and technology (S&T) roadmap for priority areas in China's modernization process. What is the purpose? And why is it? Is it a must? I think those are substantial and significant questions to start things forward.

Significance of the Research on China's S&T Roadmap to 2050

We are aware that the National Mid- and Long-term S&T Plan to 2020 has already been formed after two years' hard work by a panel of over 2000 experts and scholars brought together from all over China, chaired by Premier Wen Jiabao. This clearly shows that China has already had its S&T blueprint to 2020. Then, why did CAS conduct this research on China's S&T roadmap to 2050?

In the summer of 2007 when CAS was working out its future strategic priorities for S&T development, it realized that some issues, such as energy, must be addressed with a long-term view. As a matter of fact, some strategic researches have been conducted, over the last 15 years, on energy, but mainly on how to best use of coal, how to best exploit both domestic and international oil and gas resources, and how to develop nuclear energy in a discreet way. Renewable energy was, of course, included but only as a supplementary energy. It was not yet thought as a supporting leg for future energy development. However, greenhouse gas emissions are becoming a major world concern over

* It is adapted from a speech by President Yongxiang Lu at the first High-level Workshop on China's S&T Roadmap for Priority Areas to 2050, organized by the Chinese Academy of Sciences, in October, 2007.

the years, and how to address the global climate change has been on the agenda. In fact, what is really behind is the concern for energy structure, which makes us realize that fossil energy must be used cleanly and efficiently in order to reduce its impact on the environment. However, fossil energy is, pessimistically speaking, expected to be used up within about 100 years, or optimistically speaking, within about 200 years. Oil and gas resources may be among the first to be exhausted, and then coal resources follow. When this happens, human beings will have to refer to renewable energy as its major energy, while nuclear energy as a supplementary one. Under this situation, governments of the world are taking preparatory efforts in this regard, with Europe taking the lead and the USA shifting to take a more positive attitude, as evidenced in that: while fossil energy has been taken the best use of, renewable energy has been greatly developed, and the R&D of advanced nuclear energy has been reinforced with the objective of being eventually transformed into renewable energy. The process may last 50 to 100 years or so. Hence, many S&T problems may come around. In the field of basic research, for example, research will be conducted by physicists, chemists and biologists on the new generation of photovoltaic cell, dye-sensitized solar cells (DSC), high-efficient photochemical catalysis and storage, and efficient photosynthetic species, or high-efficient photosynthetic species produced by gene engineering which are free from land and water demands compared with food and oil crops, and can be grown on hillside, saline lands and semi-arid places, producing the energy that fits humanity. In the meantime, although the existing energy system is comparatively stable, future energy structure is likely to change into an unstable system. Presumably, dispersive energy system as well as higher-efficient direct current transmission and storage technology will be developed, so will be the safe and reliable control of network, and the capture, storage, transfer and use of CO_2, all of which involve S&T problems in almost all scientific disciplines. Therefore, it is natural that energy problems may bring out both basic and applied research, and may eventually lead to comprehensive structural changes. And this may last for 50 to 100 years or so. Taking the nuclear energy as an example, it usually takes about 20 years or more from its initial plan to key technology breakthroughs, so does the subsequent massive application and commercialization. If we lose the opportunity to make foresighted arrangements, we will be lagging far behind in the future. France has already worked out the roadmap to 2040 and 2050 respectively for the development of the 3rd and 4th generation of nuclear fission reactors, while China has not yet taken any serious actions. Under this circumstance, it is now time for CAS to take the issue seriously, for the sake of national interests, and to start conducting a foresighted research in this regard.

This strategic research covers over some dozens of areas with a long-term view. Taking agriculture as an example, our concern used to be limited only to the increased production of high-quality food grains and agricultural by-products. However, in the future, the main concern will definitely be given to the water-saving and ecological agriculture. As China is vast in territory, diversified technologies in this regard are the appropriate solutions. Animal husbandry has been used by developed countries, such as Japan and Denmark, to make bioreactor and pesticide as well. Plants have been used by Japan to make bioreactors which are safer and cost-effective than that made from animals. Potato, strawberry, tomato and the like have been bred in germ-free greenhouses, and value-added products have been made through gene transplantation technology. Agriculture in China must not only address the food demands from its one billions-plus population, but also take into consideration of the value-added agriculture by-products and the high-tech development of agriculture as well. Agriculture in the future is expected to bring out some energies and fuels needed by both industry and man's livelihood as well. Some developed countries have taken an earlier start to conduct foresighted research in this regard, while we have not yet taken sufficient consideration.

Population is another problem. It will be most likely that China's population will not drop to about 1 billion until the end of this century, given that the past mistakes of China's population policy be rectified. But the subsequent problem of ageing could only be sorted out until the next century. The current population and health policies face many challenges, such as, how to ensure that the 1.3 to 1.5 billion people enjoy fair and basic public healthcare; the necessity to develop advanced and public healthcare and treatment technologies; and the change of research priority to chronic diseases from infectious diseases, as developed countries have already started research in this regard under the increasing social and environmental change. There are many such research problems yet to be sorted out by starting from the basic research, and subsequent policies within the next 50 years are in need to be worked out.

Space and oceans provide humanity with important resources for future development. In terms of space research, the well-known Manned Spacecraft Program and China's Lunar Exploration Program will last for 20 or 25 years. But what will be the whole plan for China's space technology? What is the objective? Will it just follow the suit of developed countries? It is worth doing serious study in this regard. The present spacecraft is mainly sent into space with chemical fuel propellant rocket. Will this traditional propellant still be used in future deep space exploration? Or other new technologies such as electrical propellant, nuclear energy propellant, and solar sail technologies be developed? We haven't yet done any strategic research over these issues, not even worked out any plans. The ocean is abundant in mineral resources, oil and gas, natural gas hydrate, biological resources, energy and photo-free biological evolution, which may arise our scientific interests. At present, many countries have worked out new strategic marine plans. Russia, Canada, the USA, Sweden and Norway have centered their contention upon the North Pole, an area of strategic significance. For this, however, we have only limited plans.

The national and public security develops with time, and covers both conventional and non-conventional security. Conventional security threats only refer to foreign invasion and warfare, while, the present security threat may come out from any of the natural, man-made, external, interior, ecological, environmental, and the emerging networking (including both real and virtual) factors. The conflicts out of these must be analyzed from the perspective of human civilization, and be sorted out in a scientific manner. Efforts must be made to root out the cause of the threats, while human life must be treasured at any time.

In general, it is necessary to conduct this strategic research in view of the future development of China and mankind as well. The past 250 years' industrialization has resulted in the modernization and better-off life of less than 1 billion people, predominantly in Europe, North America, Japan and Singapore. The next 50 years' modernization drive will definitely lead to a better-off life for 2–3 billion people, including over 1 billion Chinese, doubling or tripling the economic increase over that of the past 250 years, which will, on the one hand, bring vigor and vitality to the world, and, on the other hand, inevitably challenge the limited resources and eco-environment on the earth. New development mode must be shaped so that everyone on the earth will be able to enjoy fairly the achievements of modern civilization. Achieving this requires us, in the process of China's modernization, to have a foresighted overview on the future development of world science and human civilization, and on how science and technology could serve the modernization drive. S&T roadmap for priority areas to 2050 must be worked out, and solutions to core science problems and key technology problems must be straightened out, which will eventually provide consultations for the nation's S&T decision-making.

Possibility of Working out China's S&T Roadmap to 2050

Some people held the view that science is hard to be predicted as it happens unexpectedly and mainly comes out of scientists' innovative thinking, while, technology might be predicted but at the maximum of 15 years. In my view, however, S&T foresight in some areas seems feasible. For instance, with the exhaustion of fossil energy, some smart people may think of transforming solar energy into energy-intensive biomass through improved high-efficient solar thin-film materials and devices, or even developing new substitute. As is driven by huge demands, many investments will go to this emerging area. It is, therefore, able to predict that, in the next 50 years, some breakthroughs will undoubtedly be made in the areas of renewable energy and nuclear energy as well. In terms of solar energy, for example, the improvement of photoelectric conversion efficiency and photothermal conversion efficiency will be the focus. Of course, the concrete technological solutions may be varied, for example, by changing the morphology of the surface of solar cells and through the reflection, the entire spectrum can be absorbed more efficiently; by developing multi-layer functional thin-films for transmission and absorption; or by introducing of nanotechnology and quantum control technology, etc. Quantum control research used to limit mainly to the solution to information functional materials. This is surely too narrow. In the future, this research is expected to be extended to the energy issue or energy-based basic research in cutting-edge areas.

In terms of computing science, we must be confident to forecast its future development instead of simply following suit as we used to. This is a possibility rather than wild fancies. Information scientists, physicists and biologists could be engaged in the forward-looking research. In 2007, the Nobel Physics Prize was awarded to the discovery of colossal magneto-resistance, which was, however, made some 20 years ago. Today, this technology has already been applied to hard disk store. Our conclusion made, at this stage, is that: it is possible to make long-term and unconventional S&T predictions, and so is it to work out China's S&T roadmap in view of long-term strategies, for example, by 2020 as the first step, by 2030 or 2035 as the second step, and by 2050 as the maximum.

This possibility may also apply to other areas of research. The point is to emancipate the mind and respect objective laws rather than indulging in wild fancies. We attribute our success today to the guidelines of emancipating the mind and seeking the truth from the facts set by the Third Plenary Session of the 11th Central Committee of the Communist Party of China in 1979. We must break the conventional barriers and find a way of development fitting into China's reality. The history of science tells us that discoveries and breakthroughs could only be made when you open up your mind, break the conventional barriers, and make foresighted plans. Top-down guidance on research with increased financial support and involvement of a wider range of talented scientists is not in conflict with demand-driven research and free discovery of science as well.

Necessity of CAS Research on China's S&T Roadmap to 2050

Why does CAS launch this research? As is known, CAS is the nation's highest academic institution in natural sciences. It targets at making basic, forward-looking and strategic research and playing a leading role in China's science. As such, how can it achieve this if without a foresighted view on science and technology? From the perspective of CAS, it is obligatory to think, with a global view, about what to do after the 3rd Phase of the Knowledge Innovation Program (KIP). Shall we follow the way as it used to? Or shall we, with a view of national interests, present our in-depth insights into different research disciplines, and make efforts to reform the organizational structure and system, so that the innovation capability of CAS and the nation's science and technology mission will be raised to a new height? Clearly, the latter is more positive. World science and technology develops at a lightening speed. As global economy grows, we are aware that we will be lagging far behind if without making progress, and will lose the opportunity if without making foresighted plans. S&T innovation requires us to make joint efforts, break the conventional barriers and emancipate the mind. This is also what we need for further development.

The roadmap must be targeted at the national level so that the strategic research reports will form an important part of the national long-term program. CAS may not be able to fulfill all the objectives in the reports. However, it can select what is able to do and make foresighted plans, which will eventually help shape the post-2010 research priorities of CAS and the guidelines for its future reform.

Once the long-term roadmap and its objectives are identified, system mechanism, human resources, funding and allocation should be ensured for full implementation. We will make further studies to figure out: What will happen to world innovation system within the next 30 to 50 years? Will universities, research institutions and enterprises still be included in the system? Will research institutes become grid structure? When the cutting-edge research combines basic science and high-tech and the transformative research integrates the cutting-edge research with industrialization, will that be the research trend in some disciplines? What will be the changes for personnel structure, motivation mechanism and upgrading mechanism within the innovation system? Will there be any changes for the input and structure of innovation resources? If we could have a clear mind of all the questions, make foresighted plans and then dare to try out in relevant CAS institutes, we will be able to pave a way for a more competitive and smooth development.

Social changes are without limit, so are the development of science and technology, and innovation system and management as well. CAS must keep moving ahead to make foresighted plans not only for science and technology, but also for its organizational structure, human resources, management modes, and resource structures. By doing so, CAS will keep standing at the forefront of science and playing a leading role in the national innovation system, and even, frankly speaking, taking the lead in some research disciplines in the world. This is, in fact, our purpose of conducting the strategic research on China's S&T roadmap.

Prof. Dr.-Ing. Yongxiang Lu

President of the Chinese Academy of Sciences

Preface to the Roadmaps 2050

CAS is the nation's think tank for science. Its major responsibility is to provide S&T consultations for the nation's decision-makings and to take the lead in the nation's S&T development.

In July, 2007, President Yongxiang Lu made the following remarks: "In order to carry out the Scientific Outlook of Development through innovation, further strategic research should be done to lay out a S&T roadmap for the next 20–30 years and key S&T innovation disciplines. And relevant workshops should be organized with the participation of scientists both within CAS and outside to further discuss the research priorities and objectives. We should no longer confine ourselves to the free discovery of science, the quantity and quality of scientific papers, nor should we satisfy ourselves simply with the Principal Investigators system of research. Research should be conducted to address the needs of both the nation and society, in particular, the continued growth of economy and national competitiveness, the development of social harmony, and the sustainability between man and nature. "

According to the Executive Management Committee of CAS in July, 2007, CAS strategic research on S&T roadmap for future development should be conducted to orchestrate the needs of both the nation and society, and target at the three objectives: the growth of economy and national competitiveness, the development of social harmony, and the sustainability between man and nature.

In August, 2007, President Yongxiang Lu further put it: "Strategic research requires a forward-looking view over the world, China, and science & technology in 2050. Firstly, in terms of the world in 2050, we should be able to study the perspectives of economy, society, national security, eco-environment, and science & technology, specifically in such scientific disciplines as energy, resources, population, health, information, security, eco-environment, space and oceans. And we should be aware of where the opportunities and challenges lie. Secondly, in terms of China's economy and society in 2050, we should take into consideration of factors like: objectives, methods, and scientific supports needed for economic structure, social development, energy structure, population and health, eco-environment, national security and innovation capability. Thirdly, in terms of the guidance of Scientific Outlook of Development on science and technology, it emphasizes the people's interests and development, science and technology, science and economy, science and society, science and eco-environment, science and culture, innovation and collaborative development. Fourthly, in terms of the supporting role of research in scientific development, this includes how to optimize the economic structure and boost economy, agricultural development, energy structure, resource conservation, recycling economy, knowledge-based society, harmonious coexistence between man and nature, balance of regional development, social harmony, national security, and international cooperation. Based on these, the role of CAS will be further identified."

Subsequently, CAS launched its strategic research on the roadmap for priority areas to 2050, which comes into eighteen categories including: energy, water resources, mineral resources, marine resources, oil and gas, population and health, agriculture, eco-environment, biomass resources, regional development, space, information, advanced manufacturing, advanced materials, nano-science, big science facilities, cross-disciplinary and frontier research, and national and public security. Over 300 CAS experts in science, technology, management and documentation & information, including about 60 CAS members, from over 80 CAS institutes joined this research.

Over one year's hard work, substantial progress has been made in each research group of the scientific disciplines. The strategic demands on priority areas in China's modernization drive to 2050 have been strengthened out; some core science problems and key technology problems been set forth; a relevant S&T roadmap been worked out based on China's reality; and eventually the strategic reports on China's S&T roadmap for eighteen priority areas to 2050 been formed. Under the circumstance, both the Editorial Committee and Writing Group, chaired by President Yongxiang Lu, have finalized the general report. The research reports are to be published in the form of CAS strategic research serial reports, entitled *Science and Technology Roadmap to China 2050: Strategic Reports of the Chinese Academy of Sciences.*

The unique feature of this strategic research is its use of S&T roadmap approach. S&T roadmap differs from the commonly used planning and technology foresight in that it includes science and technology needed for the future, the roadmap to reach the objectives, description of environmental changes, research needs, technology trends, and innovation and technology development. Scientific planning in the form of roadmap will have a clearer scientific objective, form closer links with the market, projects selected be more interactive and systematic, the solutions to the objective be defined, and the plan be more feasible. In addition, by drawing from both the foreign experience on roadmap research and domestic experience on strategic planning, we have formed our own ways of making S&T roadmap in priority areas as follows:

(1) Establishment of organization mechanism for strategic research on S&T roadmap for priority areas

The Editorial Committee is set up with the head of President Yongxiang Lu and the involvement of Chunli Bai, Erwei Shi, Xin Fang, Zhigang Li, Xiaoye Cao and Jiaofeng Pan. And the Writing Group was organized to take responsibility of the research and writing of the general report. CAS Bureau of Planning and Strategy, as the executive unit, coordinates the research, selects the scholars, identifies concrete steps and task requirements, sets forth research approaches, and organizes workshops and independent peer reviews of the research, in order to ensure the smooth progress of the strategic research on the S&T roadmap for priority areas.

(2) Setting up principles for the S&T roadmap for priority areas

The framework of roadmap research should be targeted at the national level, and divided into three steps as immediate-term (by 2020), mid-term (by 2030) and long-term (by 2050). It should cover the description of job requirements, objectives, specific tasks, research approaches, and highlight core science problems and key technology problems, which must be, in general, directional, strategic and feasible.

(3) Selection of expertise for strategic research on the S&T roadmap

Scholars in science policy, management, information and documentation, and chief scientists of the middle-aged and the young should be selected to form a special research group. The head of the group should be an outstanding scientist with a strategic vision, strong sense of responsibility and coordinative capability. In order to steer the research direction, chief scientists should be selected as the core members of the group to ensure that the strategic research in priority areas be based on the cutting-edge and frontier research. Information and documentation scholars should be engaged in each research group to guarantee the efficiency and systematization of the research through data collection and analysis. Science policy scholars should focus on the strategic demands and their feasibility.

(4) Organization of regular workshops at different levels

Workshops should be held as a leverage to identify concrete research steps and ensure its smooth progress. Five workshops have been organized consecutively in the following forms:

High-level Workshop on S&T Strategies. Three workshops on S&T strategies have been organized in October, 2007, December, 2007, and June, 2008, respectively, with the participation of research group heads in eighteen priority areas, chief scholars, and relevant top CAS management members. Information has been exchanged, and consensus been reached to ensure research directions. During the workshops, President Yongxiang Lu pinpointed the significance, necessity and possibility of the roadmap research, and commented on the work of each research groups, thus pushing the research forward.

Special workshops. The Editorial Committee invited science policy scholars to the special workshops to discuss the eight basic and strategic systems for China's socio-economic development. Perspectives on China's science-driven modernization to 2050 and characteristics and objectives of the eight systems have been outlined, and twenty-two strategic S&T problems affecting the modernization have been figured out.

Research group workshops. Each research group was further divided into different research teams based on different disciplines. Group discussions, team discussions and cross-team discussions were organized for further research, occasionally with the involvement of related scholars in special topic discussions. Research group workshops have been held some 70 times.

Cross-group workshops. Cross-group and cross-disciplinary workshops were organized, with the initiation by relative research groups and coordination by Bureau of Planning and Strategies, to coordinate the research in relative disciplines.

Professional workshops. These workshops were held to have the suggestions and advices of both domestic and international professionals over the development and strategies in related disciplines.

(5) Establishment of a peer review mechanism for the roadmap research

To ensure the quality of research reports and enhance coordination among different disciplines, a workshop on the peer review of strategic research on the S&T roadmap was organized by CAS Bureau of Planning and Strategy, in November, 2008, bringing together of about 30 peer review experts and 50 research group scholars. The review was made in four different categories, namely, resources and environment, strategic high-technology, bio-science & technology, and basic research. Experts listened to the reports of different research groups, commented on the general structure, what's new and existing problems, and presented their suggestions and advices. The outcomes were put in the written forms and returned to the research groups for further revisions.

(6) Establishment of a sustained mechanism for the roadmap research

To cope with the rapid change of world science and technology and national demands, a roadmap is, by nature, in need of sustained study, and should be revised once in every 3–5 years. Therefore, a panel of science policy scholars should be formed to keep a constant watch on the priority areas and key S&T problems for the nation's long-term benefits and make further study in this regard. And hopefully, more science policy scholars will be trained out of the research process.

The serial reports by CAS have their contents firmly based on China's reality while keeping the future in view. The work is a crystallization of the scholars' wisdom, written in a careful and scrupulous manner. Herewith, our sincere gratitude goes to all the scholars engaged in the research, consultation and review. It is their joint efforts and hard work that help to enable the serial reports to be published for the public within only one year.

To precisely predict the future is extremely challenging. This strategic research covered a wide range of areas and time, and adopted new research approaches. As such, the serial reports may have its deficiency due to the limit in knowledge and assessment. We, therefore, welcome timely advice and enlightening remarks from a much wider circle of scholars around the world.

The publication of the serial reports is a new start instead of the end of the strategic research. With this, we will further our research in this regard, duly release the research results, and have the roadmap revised every five years, in an effort to provide consultations to the state decision-makers in science, and give suggestions to science policy departments, research institutions, enterprises, and universities for their S&T policy-making. Raising the public awareness of science and technology is of great significance for China's modernization.

Writing Group of the General Report

February, 2009

Preface

In the present world, the status quo of the global social economy will profoundly change with the ever-changing of scientific and technological development. Faced with major historic opportunities and challenges brought about integration of the global economy, science and technology and the new technological revolution, the Party Central Committee and State Council made the major strategic decision to build the innovation-oriented country, proposed the guiding principle of adhering to independent innovation, key surmounts, supportive development and leading the future, and made the overall deployment of China's scientific and technological development, as well as grasp priorities in the process of scientific and technological development.

China' s social economy has entered into the rapid-development track: by 2020, our country will preliminarily realize the goal of building a moderately prosperous society, by 2030, China's population will reach the peak, by 2050, China will reach the level of the moderately developed countries. With the development of our society and economy, our resource demand and consumption will continuously increase, disputes between man and land as well as between economic development and ecological protection will become more and more obvious, requirements of the people for quality of the ecological environment will continuously become higher, tasks for maintaining the stable and sound ecological environment will become more and more arduous, and we will be faced with numerous challenges. Consequently, *The Outlined Plan for China's Mid- to Long-term Science and Technology Development (2006–2020)* has listed protection of ecological and environmental science as one of key areas, requiring implementation of comprehensive treatment on the regional environment and on pollution of water environment of river basins and regional atmospheric environment, technical integration and demonstration of comprehensive treatment on typical degraded zones of ecological functions, development of techniques for safeguarding the drinking water and for monitoring and pre-warning of ecological and environmental science as well as greatly improvement of technical supporting ability for the environment. However, we still fail to formulate the roadmap for scientific and technological development of ecological and environmental science (between now and 2050).

The field of ecological and environmental science is one of 18 major fields of science and technology. On the one hand, such research report is oriented to technological demands from China's social and economic development in the future 50 years; on the other hand, such report integrates the scientific research and task of the fourth-phase knowledge innovation project of CAS to shape the development strategy of the future ecological, environmental and technological fields.

From the historical point of view, the history of social development is the history of inventions. Viewing from new changes and trends in the present technological development, the shorter the innovative period, the bigger the geometric effect for economic and social development can be seen. On the one hand, the development of science and technology promoted growth of economy and society; on the other hand, demands for human civilization also serve as the driving force for scientific thinking and technical creation. Thus, while formulating the roadmap for scientific and technological development of China's ecological and environmental science, we shall think about future demands for human civilization, which will require to not only think about the evolution of the earth system (including changes in the global environment and the situation change after overlapping of artificial activities), but also carry out the forward-looking scientific thinking about the possible problems in the future economic and social development.

Major scientific discoveries will change the knowledge of mankind for the world, which is just the value of our scientific explorations. Although technical progress is conducive to the development of productivity, the double-edged sword feature of technology requires us to review side effects brought about by technical advancement. Such as, safety of genetically engineered organisms in biotechnology, disposal and resource recovery of e-wastes due to IT, bio-toxicity of nanometer particulates in nanotechnology, environmental and health effects caused by new chemicals/drugs (such as antibiotics, etc).

The research report analyzed China's social and economic development trend in the future 50 years to probe into its driving role for ecological and environmental science, define demands in fields of ecological environment and scientific technology in the future 50 years based on features and trends of ecological and environmental science in China and the world as a whole, proposed the roadmap for scientific and technological development in priority areas in China between now and 2050 by referring to the global strategy roadmaps in relevant fields, finally created the security system for implementation of roadmap strategies.

Such research report was shaped based in four symposiums of research groups and one symposium of roadmap for the scientific and technological development of ecological and environmental science in China between now and 2050. The eight established key research fields include the global climate change and its ecological process (contributed by Huijun Wang and Guirui Yu), restoration of the degraded ecological system and nursing of biodiversity (contributed by Guobin Liu and Xingguo Han), the biological geochemistry process of land/basin/coastal land (contributed by Guishan Yang), urbanization and environmental quality (contributed by Jingzhu Zhao), control and restoration of environmental pollution (contributed by Jiuhui Qu), clean production and recycling economy (contributed by Yunfa Chen), environmental pollution and health (contributed by Yongguan Zhu), advanced monitoring and forecasting techniques (contributed by Yongguan Zhu).Such report was finally concluded by Jingzhu Zhao .

In the research process of such project, we have received capital assistance and support from the Bureau of Planning and Strategy, Chinese Academy of Sciences and the help from academician Dahe Qin and researchers Xiaolei Zhang and Xi Chen of the Xinjiang Social Science Institute, as well as the great support from the Xinjiang Institute of Ecology and Geography for hosting of the project expert group meeting, and we would likely express our thanks to their care, assistance and support.

Research Group on Ecological and Environmental Strategies of the
Chinese Academy of Sciences

April, 2009

Contents

Abstract

In the future 50 years or more, with the rapid development of China's society and economy, bottleneck restrictions and stress influence of ecological and environmental science on development of economy and society become severe day by day.Faced with such formidable challenge, we must define the overall development strategy of science and technology in ecological and environmental science, make careful analysis, advanced deployment and scientific planning of the scientific and technological development of ecological and environmental science from the perspective of forward-looking, strategic and global, so as to enable the ecological environment and scientific technology to adapt to population growth, urbanization, economic development, energy and resource consumption and the global change, thereby providing greater space for China's future economic and social development.

Currently, many countries and international organizations have carried out researches into roadmaps of the scientific and technological development in some fields related to ecological and environmental science, such as, *EU's Report of the Environmental Technologies Action Plan*, *New Zealand Environmental Science Research Roadmap*, *WWF's a Roadmap for a Living Planet*, America's long-term ecology research network and strategy of national ecology observation system, such roadmaps and research strategies can serve as reference for the strategic research of the roadmap.

By 2020, China's population will be controlled within 1.45 billion, with the urbanization rate of 55%; by 2030, China's population will exceed 1.45 billion, with the urbanization rate of 65%; by 2050, China' s population will reach 1.5 billion, with the urbanization rate of over 70%.We could predict that with the growing of population, sustained rapid development of urbanization, rapid economic growth and the changing of industrial structure and consumption patterns, China will be faced with more severe environmental challenges, with problems of ecological and environmental science becoming more evident. However, in the world, among such major problems as the climatic change, species extinction and more people to be fed, many problems haven't still been resoled, moreover, such problems will put mankind's survival in danger.

The scientific and technological development of global ecological and environmental science demonstrate the following trends: ① retardation and adaptation to the global climatic change have become the focus of the scientific research of the earth system; ② the relationship between the ecological system and human well-being has become the major target of the international ecological research; ③ the research of integration of human activities, natural system and economic system has become the principal thinking of research for resolving the global ecological problems; ④ qualitative assessment and scientific prediction of the changing of ecological and environmental science have become the scientific and technological target in the ecological research; ⑤ ecosystem networking observation, simulation tests and virtual numerical simulation have become major means for comprehensive ecological problems; ⑥ environmental technology and green technology based on circulation of materials have become the a fundamental way to resolve environmental problems; ⑦ combination of the ecological system and human health has become the basic outlet for resolving regional environmental problems. In the future 50 years, technological demands driven by the development of society, science and technology include:

① to provide science and technology support for the ecological civilization and eco-engineering construction; ② to provide science services for management and optimized distribution of natural resources and for sustained regional development; ③ to the scientific and technological requirements for coping with the global climatic change and the global economic globalization; ④ to provide science and technology support for fulfilling international conventions and international cooperation; ⑤ to accumulate the scientific data of the long-term spatialization to promote the scientific development of the earth system; ⑥ to safeguard the environmental safety; ⑦ to maintain the health of ecosystem.

Based on the existing scientific and technological development planning, and according to technological demands driven by the development of economy, society, science and technology, we proposed eight key research fields of ecological and environmental science:

① global climatic changing and its ecological process; ② restoration of the degraded ecological system/nurturing of biodiversity; ③ urbanization and environmental quality; ④ the biological geochemistry process of lands/river basins/coastal zones; ⑤ control and restoration of the environmental pollution; ⑥ clean production and recycling economy; ⑦ environmental pollution and health effect; ⑧ advanced monitoring and forecasting techniques. Common technical approaches of the eight major key areas include: ① multidisciplinary comprehensive development of innovative and high technologies; ② building of ecological and environmental science observation platform and system theory; ③ realization of data integration and system simulation.

Goals for strategic implementation of the roadmap for the scientific and technological development of China's ecological and environmental science are as follows:

Around 2020, the trend of degradation of China's ecological and environmental science will be basically checked; around 2030, recovery of the typical degenerated ecological system and restoration of polluted environment; around 2050, the beautiful environment and healthy ecology will be realized.

The roadmap strategy for scientific and technological development of China's ecological and environmental science will be implemented by establishing the security system in the following six aspects: ① to create the large research plan aimed for the methodology research; ② to set up the basic research platform for the problem-oriented interdisciplinary ecological and environmental science; ③ to build 2–3 optimized multiple-target comprehensive experiment and demonstration areas (cities); ④ to set up the pilot center integrating of state integrated ecological environment monitoring, data integration and system simulation; ⑤ to join with the neighboring countries to establish the Asian center for the global change.

1 Significance of Research on the Roadmap for the Development of Ecological and Environmental Technology

In the next 50 years and beyond, with the rapid economic development in China, bottleneck restrictions and impacts on ecosystems and environment will become increasingly serious with each passing day. Faced with such a severe challenge, we must thoroughly apply Deng Xiaoping's Theory, the important thought of Three Represents, the spirit of the 17th National Congress of CPC, take the scientific outlook on development as the guiding principle, define the general strategy of the future scientific and technological development of ecological and environmental science. To fulfill that, we must implement the *Outlined Plan for China's Mid- to Long-term Science and Technology Development (2006–2020),* Medium- to Long-term Scientific and Technological Development Plan of Chinese Academy of Sciences in the field of Resources Environment, make careful analysis of recent developments in ecological and environmental science, so as to enable science and technology to improve ecosystem safety and environmental quality. This will ensure that our scientific community to adapt to requirements for economic development, population growth, urbanization, new patterns of energy and resource consumption and the global changing, thereby providing theoretical and technical support for the sustainable development of China's economy and society in the future.

Significance of the strategic research of the roadmap for the scientific and technological development of ecological and environmental science is mainly embodied in the following aspects:

1.1 Safeguard National Ecological System and Environmental Security

Inherent vulnerabilities of the natural environment in most regions of our country, heavy population pressure, rapid economic development with an irrational development mode have all contributed to the serious degradation of our ecological system, continuous aggravation of the environmental pollution and severer environmental health problems. Ecological and environmental problems are further exacerbated by global change. Confronted with these problems, we must develop science-based approaches to the protection of ecosystems from the roots, restore the degraded ecosystems, control and remediate the polluted environments, prevent the generation of new pollution, and minimize harms of pollution imposed on ecosystems and human health, so as to actively cope with impacts of the global change on the earth system and to safeguard the national ecosystem and environment safety.

OECD's Evaluation on Environmental Performance of China

Since the reform and opening up of 1978, China has made great achievements in the economic development. Over the past 15 years, economy has been kept growing at an annual average rate of 10.1%. China is now the world's fourth largest economy. Substantial overseas investments and the ever-increasing market-driven forces have promoted China's integration into the global economy. In this process, large amount of people have come out of poverty. However, China has a population of 20% of the world population, and its per capita GDP is relatively low (according to the standard for evaluating the purchasing power, China's per capita GDP is 6000 US dollars, while that of OECD countries is 25000 dollars). With the contrast between the affluent coastal provinces to less developed western provinces, geographical disparities between poor and rich are relatively bigger. Migration of massive people has brought about the rapid urbanization process (the present urbanization rate is 43%), with a widening income gap between rural and urban areas. Poverty remains the severe challenge facing Chinese rural areas. Rapid economic development, industrialization and urbanization have brought great pressure to the environment and caused harms on human health. Having been clearly aware of degradation of the environment, the Chinese government is adopting a balanced development mode, proposing concepts of "a harmonious society" and "scientific outlook on development". Countermeasures are applied to formulate the development plan of the national economy and society (including intensifying of environmental management), draw up the modernized environmental law, strengthen construction of environmental agencies, give top priority to management of environmental and natural resources.

Although, atmospheric pollution in some China's cities is still on the list of world's worst cities, with energy consumption 20% higher that the average level of OECD countries and 1/3 river basins seriously polluted. Waste management, desertification control and natural and biological diversity are all faced with challenges. Environmental pressure and demands for energy and other resources brought about by the rapid development of China's economy highlights the environmental sustainability issue due to the present global production and consumption patterns.

In terms of the atmospheric environment, in 2005, the air quality in about 60% of the cities at the county level or above reached the state II-class environmental quality standard, meeting the 50% goal set forth in the Tenth Five-Year Plan. After falling steadily since the 1990s, ambient sulphur dioxide concentration in urban areas began to be on the increase once again in 2002. In 2000–2004, emissions of the fixed source sulfur dioxide around the country increased 13%, failing to meet the goal of reducing 10% set forth in the Tenth Five-Year Plan. In the air pollution control district, sulfur dioxide (SO_2) emissions reduced 2%, but not 20% stipulated in the plan. Likely, the proportion of cities seriously affected by acid rain was on a lower level of 2% in 2000, but increased 10% in 2004. The goal for reducing emissions of pollutants is not enough to meet requirements of the standard for atmosphere environment quality. Up to now, we fail to pay due attention to controlling of nonvolatile organic compounds (VOCs) and poisonous atmospheric pollutants. Laws, regulations and permits concerning control of air pollution weren't effectively implemented. The energy consumption intensity for China's unit GDP is 20% higher than the average level of OECD countries, which began to rise again in 2001 after the falling at the initial period of performance evaluation. The Chinese government has given priority to reducing energy consumption for the economic development.

Double supply of crude energy is predicted to meet the requirement of the quadruple growth of GDP by 2020 (compared to that of 2000). However, in 2006–2010, it will be a great challenge to reduce 20% of the energy consumption. Although there are many potential benefits in many aspects (such as, emission reduction of air pollutants and greenhouse gases as well as improvements in independence and safety of energy and economic efficiency).China failed to realize the coal washing rate of 50%, with the proportion of installed flue gas desulfurization equipments below expectations. Although car owners are still the fewer, the number of motor vehicles doubled in 1995–2000, consequently, motor vehicles pollution has become the primary source of urban air pollution; urban transportation efficiency is gradually falling, and urban large-capacity transportation is not attached enough importance to, utilization of bicycles is in downtrend momentum; quality of fuel oil is poor (such as higher sulphur content).

Water environment conditions in China should be given top priority to. Firstly, due to emissions of industrial, agricultural and domestic pollutants, part of rivers, lakes and coastal waters were seriously polluted. Pollution has caused severe degradation of the water system, constituting the major threat on the human health and economic growth. Utilization of raw water affected the economic and social development, especially in those poverty-stricken and less developed regions, massive investments should be encouraged for water services: ① implementing investment reserves in urban areas to meet demands from mass migrant rural people; ② while conducting investment in rural areas, affordability should be considered; ③ in the less developed regions, forms of development assistance or transfer should be applied. Secondly, occupancy volume of China's per capita water resources is extremely low (a quarter of the global average level), and distributed unevenly (for example, occupancy volume in north and west areas only accounts for 1/10), in more than 600 cities, 400 cities are faced with shortage of water resources. Thirdly, with worsening of pollution and shortage of surface waters, demands in many rural and urban areas for the underground water are far beyond the re-supply rate. It is impossible to maintain efficient or inefficient water consumption in urban and rural areas. Our country is implementing a huge project, that's to transfer 40 billion cubic meters of water to north plain area from the Yangtze valley in 2020.However, if failing to make strict control of demands from urban, industrial and rural users and if without the sustainable usage of water resources, the water diversion project still can't meet needs for economic growth and ecological restoration. Finally, about 70% of China's water resources were applied to agriculture, in which the irrigated area accounts for 40% of the acreage under cultivation. Agricultural development and rural residents (lacking of sewage systems) serve as major sources of pollution. In order to ensure sustained water management, agricultural water consumption should be reduced, with the diffused pollution identified and prevented.

Ability for treatment and disposal of wastes needs to be improved. Many wastes were stored in designated places for treatment (for example, nearly 50% of urban wastes), in addition, part of wastes were dumped in an uncontrolled manner. More wastes were piled at random in the urban surrounding areas, brining risks to the human health and environment. The goal of the urban refuse landfill amount of 150kg/d set forth in the Tenth Five-year Plan wasn't achieved. During the period of assessment, total output of wastes increased 80%.Compared to air and water environment management, wastes management wasn't paid enough attention to in terms of allocation of investment funds of national governments. Waste charging is far below the operating cost of wastes management, and local governments are confronted with trouble in levying. On the whole, refuse landfill was exaggerated (accounting for 44% of urban ultimate wastes), effects of separate collection and cycling usage of resources are not ideal. Incineration and composting only accounted for 3% and 5% in urban waste treatment respectively. Responsibilities for wastes management are dispersed in many institutions. Law enforcement is not effective enough to make a distinction between large, medium and small enterprises.

Source of data: OECD. OECD's Environmental Performance Review of China - conclusions and recommendations (final draft). 2007. (OECD) Environmental Performance Working Conference

1.2 Promote Reasonable Urbanization and Regional Sustainable Development

Urbanization has thoroughly changed the basic structure and evolutionary process of the man-land relationship and the land system. Currently, blind urbanization and out-of-order regional development are the two cruxes to aggravate disputes between mankind and the nature and to impose restrictions on China's modernization drive.

The strategic implementation of sustained regional development will exert profound influence on the new pattern for development and application of a land surface representing the ecological civilization, while implementation of the scientific outlook on development and building of a harmonious society are crying for establishing of a powerful tech-aid system in fields of urbanization and sustained regional development. Rational selection of urbanization paths and sustained regional development require a good grasp on conditions of ecological and environmental science and on their relations with society and economy, and a scientific planning.

Eight Major Problems in China's Sustainable Development

I. The macro-coordination ability for the environmental protection and sustainable development slightly decreased, including the lower status of State Environmental Protection Administration.

II. In spite of the advanced concept of new-type industrialization, there is a lack of concrete channels for realization of industrialization.

III. The route of sustainable urbanization hasn't established, with resources-environment protection failing to integrate with the mainstream of urban development.

IV. In terms of energy-saving and emission-reduction, efficiency index is realistic, however, in terms of emission reductions, high cost will be paid for such a total amount index.

V. Decision-making of many polices wasn't based on complete basic data and information, and the process to determine policy objectives and make decisions is rudimentary, and lacks truly independent argument system and rational procedures, thereby causing limited capital failing to realize priority policy objectives.

VI. The long-term planning is lack in ecological protection and construction engineering, restoration and protection of ecology fail to be combined with peasants' long-tem alternative livelihood, consequently, when the project was completed, trees in returned forests are still in the risk of being cut down and reclaimed. In addition, fairly high allowance standards, inclination to the departmental benefit, overlapping of project administrative organizations and other problems all affected effectiveness and sustainability of the project and policy.

VII. Basic institution arrangements and policy guarantees for realization of the sustainable development and construction of conservation-minded society aren't effectively enough, including faultiness of the property system, price mechanism and economic incentive policies, all contributed to difficulties in generating motivation to save resources and reduce emission pressure.

VIII. Disproportion between the nature of trans-departmental and cross administrative division of resources and environment and the administrative jurisdiction made the present institutional arrangements and policy measures hardly carry out its functions.

Source of data: Strategic research group of the sustainable development, Chinese Academy of Sciences. 2008. China Sustainable Development Strategy Report - Policy Review and Outlook. Beijing: Science Press

Constraints of Resources and Environment in China's Sustainable Development

i Constraint of resources shortage

China's large population have imposed great pressure on resource and environment, especially in the mass consumption period, the contradiction between supply and demand of resources is growing more acute.

The first is the energy problem. With the rapid growth of China's economy and continuous acceleration of urbanization and industrialization, energy needs are increasing day by day. In 2007, the yearly total energy consumption was 2.65 billion tce, a 7.8% increase over the previous year. Consumption of crude oil was 340 million tons and increased 6.3%, only next to the United States, which is one of consumer countries. Owing to the extensive economic growth, irrational energy structure, low-level technical equipment of energy, relatively backward managerial level and high-level pricing of energy resources, especially higher dependency on importing of petroleum, energy risk and safety are not only economic problems, but heavily intertwined with political and military affairs.

The second is the problem of water resources. Since the end of World War II, world population increased from 2.5 billion in 1945 to 6.5 billion in 2007. Rapid population growth caused the dramatic increasing of mankind's needs for water resources. As one of the most populous countries with relatively lack water resources, China's per capita water capacity accounting for 1/4 of the world's average level, low utilization efficiency and serious pollution have become one of major bottlenecks to restrict its economic and social development.

The third is land, especially the arable lands problem. The world's desert area is about 33.7 million square kilometers, accounting for 1/4 of the land area, with nearly 500 million people living in the desert area. China's desertification area is about 2 million square kilometers, which accounts for 27.5% of the surface area and spreads at the speed of 2460 square kilometers.

In 2007, China's arable land area was 1.826 billion acres (one acre is converted into 667 square meters), accounting for 10% of the surface area, its per capita arable land area is less than 1/2 of the world's average level. With the continuous expansion of construction scale, the red line of 18 acres of arable lands is likely to be broken.

The fourth is food security. In 2007, China's grain output reached 501.5 billion kilogram, it has been a great miracle to achieve the basic grain self-sufficiency of 1.3 billion population. In recent years, with the enormous rising of world grain price, new changes have taken place in the supply-demand relation. Globally, 100 million population is faced with famine. In our agricultural production, especially in grain production, owing to the ever-decreasing of arable lands, abnormal climatic changes and aggravating of pollution, the security risk of grain for survival of the future 1.5 billion population will exist for a long time. Facts have proved that China has had no choice but to realize the basic self-sufficiency of grain on its own.

ii Restrictions from the environment supporting capacity

As a developing country with less developed industrialization, China is faced with more server problems such as ecological damage, environmental pollution, relatively lack resources and large population.

First, the general deterioration trend of ecological environment hasn't been fundamentally reversed. China is one of countries suffering the most serious natural disasters. Natural disasters happen frequently, with a wide range of distribution, numerous varieties and strong destructiveness. Unexpected and common disasters such as earthquakes, typhoons, drought, red tides, forest fires, sandstorm and plant disease and pests have caused heavy losses of life and property of the people. In particular, the heavy losses caused by the outbreak of snow and ice storms in South China in the early half of 2008 and Wenchuan Earthquake disaster are worth thinking deeply. Owing to reductions of forests (the annual reduction of forest resource is about 100 million cubic meters), grass- and deterioration (accounting for 1/5 of the grassland area), serious soil erosion (the annual erosion of silt is about 5 billion tons), expansion of land desertification, unsuitable human. Human settlement areas have accounted for nearly 1/3 of the land area. Second, vulnerability of state ecosystem natural cover areas is nearly approaching the utmost limit. State ecosystem natural cover areas are headwaters areas of China's great rivers, also ecological safety regions for the survival and development of the Chinese nation. Such vast areas are stretching from the Qinghai-Tibet plateau in the west, and from the Mongolia plateau and Qinghai-Tibet plateau to the Da Hinggan Ling and Xiao Hinggan Ling in the east, where the ecosystem is getting worse and very vulnerable, forests and wetlands are reducing, and the people living a poverty life on the whole. Facts have proved that once the ecosystem was damaged, the period for recovery will be extremely prolonged. According to "Environmental Sustainability Index" issued by Swiss in 2005, China is ranked the last 14th among 146 nations and regions in the world. Third, the climatic change has brought about more complicated and elusive frequency, intensity and regional distribution of extreme meteorological disasters in China. Focusing on the problem of how to remediate the climatic change, diplomatic negotiations will become more intense in the global arena, and as a result, China is faced with greater pressure for reducing emissions of greenhouse gases. The climatic change isn't pure an issue of climate and environment. The climatic has influenced human's survival and development, profoundly involving agricultural and grain security, ecological security and public sanitation security.

Source of data: Zhang Weiqing.2008. Thinking about sustainable development of China. CPPCC, (3): 46–49

1.3 Provide Scientific and Technological Support for the National Responding System of Public Emergency

In January, 2005, the state council issued *Overall Emergency Preplan for National Sudden Public Incidents*, which incorporated land resources security, food security and major natural disasters into the public security field, and gave express orders that Chinese Academy of Sciences should provide scientific and technological support for strengthening studies on techniques for public security monitoring, forecasting, pre-warning and emergency disposal, so as to establish and improve the emergency technological platform of public security.

Coping with sudden public incidents requires prompt and exact dynamic information about resources environment, however, objectivity and accuracy of such information are difficult to be evaluated, which are often behind in time, once sudden incidents occurring, national and public interests will be severely damaged. Thus, our most urgent work is to monitor and evaluate important information about resources environment by using of the earth observing technology and through integrated analysis of multi-disciplinary and multi-source information, build the remote sensing monitoring and early warning platform of resources and environment and emergency response service system, improve the national ability to foresee and handle major problems, so as to provide scientific and technological support for the national public security system.

1.4 Fulfilling the Obligations of International Environmental Conventions and the Requirements of National Environmental Diplomacy

Currently, China has signed up a series of international conventions including *UN Framework Convention on Climate Change, United Nations Convention to Combat Desertification, Convention on Biological Diversity and Stockholm Convention on Persistent Organic Pollutants*. During the process of international activities and honoring conventions in such environmental fields, the state urgently needs to take full and detailed scientific data and research results as the scientific and technological support, and especially needs to make studies on key scientific problems related to conventions, provide accurate scientific information, countermeasures and plans for building China's good international image, so as to create a sound environment for social and economic development; meanwhile, fulfilling of international conventions is also conducive to improving China's managerial level of ecological and environmental science and to promoting international cooperation.

Brief Introduction to *UN Framework Convention on Climate Change*

In the FWCC of 1979, climatic change was put on agenda as an issue attracting attention from the international community. With the gradual deepening understanding on the global climate and ever-increasing of carbon dioxide concentration in the atmosphere, UNEP and WMO set up the Intergovernmental Panel on Climate Change (IPCC) in 1988.

In 1990, IPCC published the first assessment report on climate change. Such report provided scientific basis for the climatic changing. Based on the report issued by IPCC, the UN general assembly set up the Intergovernmental Negotiating Committee in 1990 to engage in negotiations on *the UN Framework Convention on Climate Change* (UNFCCC).

In the UNCED (namely Rio Earth Summit) of 1992, 154 countries signed

UNFCCC, aiming to concentrate greenhouse gases in the atmosphere at a level, at which dangerous climatic changes due to human activities can be avoided. Within a certain time period, so as to enable ecosystems to adapt to climatic changes and prevent grain production from being endangered, as well as to promote the sustainable development of economy.

The convention classified countries into two groups:the developed countries which should be mainly responsible for the emissions of green-gases (usually named appendix I countries), and the developing countries which (usually named as non-appendix I countries) which will increase their proportion in man-made emissions in future.

Signatory countries of this convention are obliged to compile national summary list of emission sources of greenhouse gases. They also promised to formulate state strategies to adapt to and relieve climatic changes, and incorporate climatic change considerations into formulating of policies related to social economy and environment. They must also improve functions for sustainable management, energy-saving and for strengthening of greenhouse gas sinks, including forests and all other land, coastal and marine ecosystems.

This convention has set up a financing mechanism to provide developing countries with donations and preferential loans, so as to assist them in fulfilling conventions and handling climatic changes. This convention designated Global Environment Facility (GEF) as its temporary financing mechanism, passed the memorandum of understanding identical with GEF at the second contracting parties' conference in 1996, and specified their respective responsibilities and obligations. In 1998, the fourth contracting parties' conference entrusted GEF as its permanent financing mechanism and institution, reviewing it every four years. The financing mechanism is responsible for the contracting parties' conference, the latter is in charge of determining climatic changing policy, planned priority areas and standards for obtaining assistance, thus, the contracting parties' conference regularly provide policy guidance to the financing mechanism.

Source of data: http://www.mep.gov.cn/inte/lydt/200302/t20030221_84432.htm

Brief Introduction to *United Nations Convention to Combat Desertification*

i General information

United Nations Convention to Combat Desertification is one of three major environment-related international treaties under the framework of United Nations Conference on Environment and Development (*The 21 Century Agenda*).This convention was passed at the Diplomatic Conference on June 17, 1994 in Paris, France and took into effect on December 26, 1996.

ii Purpose

The convention is designed to combat desertification and relieve the drought impact in countries suffering severe drought and/or desertification, especially in

Africa, so as to assist affected countries and regions to achieve the sustainable development.

iii Membership

By April 26, 2005, 191 countries were approved to join in the Convention.

iv Major principals

H. E. Hailia Arba Diallo, executive secretary, the former foreign minister of Burkina Faso, has served as UNDP official, ambassador in China, head of Burkina Faso Permanent Mission to the United Nations, and in 1992, he was in charge of the transitional secretariat after the United Nations Conference on Environment and Development.

v Headquarters

The permanent secretariat is located in Bonn Germany. Website: http://www.unccd.int

vi The financing mechanism

(1) Global Mechanism (OM);

(2) Global Environment Facility (GEF)

vii Organizational institution

(1) The Contracting Parties' Conference of the Convention: the supreme decision-making body of the Convention.

(2) Committee on Science and Technology: to provide suggestions and information of science and technology to the contracting parties' conference.

(3) Committee for the Review of the Implementation of the Convention: it was set in the fifth contracting parties' conference of Convention in 2001, in charge of supervising on and urging contracting countries to fulfill the convention.

(4) Permanent Secretary: as the executive agency, it is in charge of conference arrangement, preparation of meeting documents and coordination of relations between internal and external institutions of this Convention.

In 1997–2001, one session of the Contracting Parties' Conference of the Convention was held every year. Since 2002, a session of the contracting parties' conference of the Convention has been held every two year. Committee for the Review of the Implementation of the Convention holds a session every year.

Between August 25 and September 5 of 2003, the sixth Contracting Parties' Conference of this Convention was held in Havana, capital of Cuba. Of which, September 1 and 2, Special High Level Segment was held, at which ministerial-level senior officials of countries evaluated fulfillment of the Convention and expressed their positions. Two affiliated organizations of this Convention – Committee for the Review of the Implementation of the Convention and Scientific Committee also held their sessions during the conference.

Key issues of the conference include: report on the first session of Committee for the Review of the Implementation of the Convention (CRICI) and suggestions from the Scientific Committee; reviewing of budgets between 2004 and 2005; discussions of World Summit on Sustainable Development and achievements related to the Convention.

The Chinese delegation headed by Lieke Zhu, deputy director-general of the State Forestry Bureau attended the conference. Lieke Zhu, head of Chinese delegation introduced China's measures and achievements in combating desertification, expressing that China will intensify cooperation with other countries around the world. China affirmed the work carried out by the Committee for the Review of the Implementation of the Convention while urging developed countries to sincerely fulfill its obligations, and requiring to coordinate well the work of the current financing mechanism – GM and GEF. In terms of budgeting problems, China advocates that the Contracting Parties' Conference shall promptly make decisions on problems such as expenditures, so as to give full play to the role of the Regional Coordinating Unit (RCU).China supports to adopt effective measures to fulfill World Summit on Sustainable Development and fruits related to the Convention.

The third session of Committee for the Review of the Implementation of the Convention was held in between May 2 and 11 of 2005 in in Bonn, Germany.

China signed the Convention on October 14, 1994, with the approval letter was deposited on February 18, 1997. The convention took into effect in May 9, 1997 in China.

Source of data: http://world.people.com.cn/GB/8212/60991/60996/4269372.html.

Historical Background of the *Convention on Biological Diversity*

In 1972, United Nations Conference on human environment was held in Stockholm, in which United Nations Environment Programs was set up, governments of the world signed numerous regional and international treaties to deal with such issues as protection of wetlands, administration of trading in international endangered species, such treaties and relevant agreements for regulating toxic chemicals pollution worked together to slow down the trend of deteriorating environment, although such trend wasn't basically reversed, international bans and restrictions on hunting, excavation and reselling of some animals and plants have reduced behaviors of over-hunting and poaching.

In 1987, World Commission on Environment and Development (Brundtland Committee) concluded, if wanting to develop economy, we must reduce damages on the environment, such epoch-making report was titled with "Our Common Future", indicating that mankind has possessed the sustainable development ability to satisfy their own needs without sacrificing interests of later generations, meanwhile, such report called on "a new epoch characterized by a sound and green economic development.

In 1992, the largest-scale United Nations Conference on Environment and Development was held in Rio de Janeiro, Brazil. World leaders signed a series of protocols with historic significance on this summit, including two binding agreements: *Convention on Climate Change and Convention on Biological Diversity*. The former is aimed for controlling emissions of greenhouse-effect gases such as industrial and other greenhouse gases including carbon dioxide, the latter is the first global agreement for protection of biological diversity and the sustainable utilization. *Convention on Biological Diversity* soon prevailed, more than 150 countries signed the document at the Rio Conference, hereafter, a total of 175 countries approved the agreement.

Three principal objectives of the Convention:

(1) Protection of biological diversity

(2) Sustainable utilization of components of biological diversity;

(3) Sharing in commercial interests of genetic resources in a fair and just manner and utilization in diversified forms.

The convention involves a wide range of objectives, and aimed to deals with many major issues about the mankind's future, serving as the milestone of the international law. The convention reached the common ground that protection of biological diversity is in the common interests of the mankind and an indispensable part in the development process. The convention covers all ecological systems, species and genetic resources and integrates traditional protection efforts with the economic goal for the sustainable utilization of biological resources, thereby establishing the philosophy for sharing genetic resources in a fair and reasonable manner, especially from the commercial perspective, the convention involved rapidly developed biotechnology, including development, transfer and beneficial sharing of biotechnology and bio-safety. More importantly, the convention possesses legal binding force and the contracting parties are obliged to exercise its terms.

The convention reminds the decision makers that natural resources aren't endless, thereby providing a brand-new concept – the sustainable utilization of biological diversity, past efforts are focused on protecting of some special species and habitats. The convention thinks that biological systems, species and genes must be used to serve the common interest of mankind, which can't be obtained in the utilization mode and at the utilization speed likely to cause long-term degradation of biological diversity.

Based on the precautionary principle, the convention provided a guide for deciders: while biological diversity obviously decreasing or falling, we can't use the excuse of lacking in sufficient scientific conclusions as measures to be adopted to reduce or avoid such threat. The convention confirms that protection of biological diversity requires substantive investments, however, meanwhile, it emphasizes that protection of biological diversity should bring obvious returns to our environment, economy and society.

Subjects of the convention include:

(1) Measures and incentive means for protection and sustainable application of biological diversity;

(2) Acquiring of genetic resources;

(3) Technical acquisition and transfer including biotechnology;

(4) Cooperation in scientific and technological fields;

(5) Impact evaluation;

(6) Education and public awareness;

(7) Capital source;

(8) State report on fulfilling of Convention obligations.

Convention obligations to be fulfilled: as an international treaty, the convention identified common difficulties, and set the complete goal, policy and universal obligations while carrying out technical and fiscal cooperation, however, the contracting parties themselves are largely responsible for achieving the goal.

Private companies, land owners, fishers and farmers engaged in numerous activities affecting the biological diversity, the government should fulfill its leadership responsibilities to formulate measures for guiding them in protecting state-owned lands, and biological diversity of waters by using of laws and regulations related to natural resources. According to the Convention, the government will assume responsibility for protection and sustainable utilization of biological diversity, and must develop the national strategy and action plan for biological diversity, as well as incorporate such strategy and plan into the extensive plan for state environment and development, which is of particular importance to forestry, agriculture, fishery, energy, transportation and urban planning. Other obligations of the convention include:

(1) To identify and monitor important components of the biological diversity to be protected;

(2) To set up reserve areas for protection of biodiversity and promote regional development in an environmental friendly manner;

(3) To cooperate with local residents in recovering and restoration of ecosystems and in promoting species restoration;

(4) Participated by local residents and communities, respect, protect and maintain traditional knowledge for the sustainable utilization of biological diversity;

(5) To prevent introducing of foreign species threatening ecological systems, habitats and species and to control and eliminate such species;

(6) To control risks caused by organisms modified through modern biotechnology;

(7) To promote the public's participation, especially in evaluating environmental influence caused by development projects threatening biological diversity;

(8) To improve public awareness of importance of biological diversity and necessity for protection of such diversity;

(9) To report to the contracting parties on how to achieve the goal of biological diversity.

Institutions of the convention:

The highest governing body of the convention is the conference of contracting party (COP), which is composed of governments ratifying the convention (including the regional economic integration organization), and in charge of inspecting the progress of the convention, determining priority protection focuses for member states and formulating the work plan. COP may revise the Convention, set up consultants and experts group, review the progress report submitted by member states and carry out cooperation with other organizations and conventions.

COP may obtain professional knowledge and support from other organizations set through the Convention, such as:

SBSTTA. It is composed of experts from relevant fields in member states, playing a key role in providing advices for scientific and technical issues.

Data exchange mechanism. It is the Internet-based network designed for promoting and technological cooperation as well as information sharing. Secretariat is set in Montreal, Canada, and closely connected to the UNEP. It is mainly designed to organize meetings, draw up documents, assist member states in fulfilling the work plan and cooperate with other international organizations, as well as to gather and provide information. Where appropriate, COP can establish specialized commissions or mechanisms, in 1996–1997, COP established the Working Group on Biosafety and Work Group on Indigenous Knowledge and Local Communities.

Global Environment Facility (GEF). While carrying out activities related to the convention, developing countries can obtain assistance via convention's financing mechanisms such as GEF.GEF project jointly sponsored by UNEP, UNDP and the World Bank promoted international cooperation and gave assistance in four fields exerting great influence on the global environment: loss of biodiversity, climatic change, deletion of the ozone layer and exhaustion of international water resources. By the end of 1999, GEF has provided assistance valued at 1 billion dollars for biodiversity projects of 120 countries.

Source of data: http://www.biodiv.gov.cn/dyxgy/200402/t20040226_88571.htm

Brief Introduction to *Stockholm Convention on Persistent Organic Pollutants*

Between May 22 and 23 of 2001, UNEP organized the convocation of the diplomacy plenipotentiary conference of *Stockholm Convention on Persistent Organic Pollutants*, in which the convention was finally passed, the convention is the third approved international treaty designed to protect the global environment by globally reducing emissions, an important step taken by the international community to impose priority control action on toxic chemicals. Currently, at least 156 countries have signed the convention. The Chinese government signed the convention on May 23, 2001.Such convention was aimed to reduce or eliminate emissions of persistent organic pollutants, so as to protect human health and the ecological environment from their harms. The first batch of controlled chemical substances published by the convention include three categories covering 12 varieties: (1) pesticides: DDT, chlordane, Mirex, ;aldrin, Endrin, heptachlor and HCB; (2) industrial chemicals: PCBs; (3) side products: furan.

Persistent organic pollutants refer to chemical substances which permanently exist in the environment, accumulate through the food web, and exert a negative impact on human health and environment. Different from conventional pollutants, persistent organic pollutants are hard to be degraded in the natural environment, and can be transferred via such carriers as water or air. After accumulating in the human body or animal, such persistent organic pollutants can easily bring cancer risks to human health. Many persistent organic pollutants not only posses carcinogenicity, mutagenicity and teratogenicity, but also disturbing the internal secretion. Studies indicate that influence of persistent organic pollutants on mankind will last as the generations pass by, and constitute a major threat on humans' survival, reproduction and sustainable development.

Source of data: http://www.cneac.com

2 Overview of Researches of Roadmaps and Plans for the Scientific and Technological Development in Relevant Fields in Foreign Countries

Technology roadmap is an effective means for innovative management. It helps users to understand the development orientation of such field and key techniques for realization of objectives. Since technology roadmap was firstly applied in enterprises in the 1970s and 1980s, it has been widely utilized and developed in the scientific research as an innovative management tool. Research and drawing of technology roadmap are conducive to identifying the existing national ability of science and technology and the bottleneck that affects the future development. As a major basis for decision-making on planned investment and resource allocation, studies on roadmaps for the scientific and technological development focus on state priorities areas, promoting of scientific and technological innovation and the future consideration of the sustainable development of the national science and technology and social economy. Currently, many countries and international organizations have carried out researches of roadmaps for the scientific and technological development in fields related to ecological and environmental science, which can serve as references for strategic research of the roadmap.

Generation and Role of Technology Roadmap

Roadmap is the diagrammatic presentation for directing people to reach their desirable destinations. Such general definition grasps the nature of roadmap: to shape the common sense of the future and their future desirable goal in a team. Drawing of roadmap is just the learning process of the team, and they may find differences and new orientations during this process. Drawing of roadmap is conducive to improving the accuracy and efficiency in team's development of their internal language and communications among team members. The roadmap is also helpful for communications among team members, other people (such as customers and suppliers) and other teams in terms of future assumptions and plans.

Like as other common roadmaps, technology roadmaps also boast many similar purposes. For example, technology roadmaps can be used to depict evolutions of past or future science or technology along some possible directions and the future development path of certain industry.

In enterprises, demands for technology roadmap are generally embodied in such aspects as the level of product line, definition of product development, business integrating strategy, product features, integrating cost and technological strategic targets. Technology roadmap integrated requirements of enterprise development and technology evolution.

In the 1970s and 1980s, Motorola and Corning firstly drew technology roadmap. Corning advocated to draw based on key events to reach the overall business strategy and business unit strategy; Motorola adopted technology evolution and localization method. Motorola's methods are more seen in practice of America's technology management. Under the leadership of CEO Robert Galvin of that time, Motorola launched the action of drawing the technology roadmap throughout the corporation, which was designed to encourage business managers to appropriately focus on the future technology and provide them with a tool for forecasting the future process. Such method was introduced to help balance long and short-term subjects, strategies and operated affairs in technological and other fields. In the paper edited jointly by them, Willyard and McClees think that Motorola was the first one to apply roadmap and method in this respect. According to the paper, drawing of roadmap is considered to provide channels for exchanges between engineers of design, development and research and there fellows in charge of market research and marketing, with technology designed to serve future product development and application. Publication of Motorola model and experience arose tremendous interest from top-level managers in enterprises, which directly led to application of technology roadmap by Rockwell Motor Company in 1995.

Source of data: Liu Haibo, Li Ping. 2004. Generation and role of technology roadmap. Science & Culture Science-Tech Waves, (9): 8–9

2.1 Environmental Technologies Action of EU

Environmental Technologies Action Plan was formulated by the European Commission on January 28, 2004, which was designed to reduce pressure on natural resources to the greatest extent, improve the quality of life of European citizens and stimulate the economic growth. Such is a perfect supplement to European Sustainable Development Strategy, consistent with Lisbon strategy and beneficial to developing countries. EU encouraged application of environmental technology in various forms of investment and made relevant objective decision making, all these measures will be conducive to broadening the market and reducing costs, which should be realized through joint efforts made by EU commission, member states or research groups, organizations, industries and the social public.

Environmental Technologies Action Plan was designed to eliminate obstacles to develop environmental technology in an all-round way, protect the environment and promote contestability and economic growth, and to ensure Eu's leading role among developing countries and application of environmental technology, as well as to mobilize every effort to support such plan. In order to effectively implement such action plan, EU has studied the environmental technology in such three key: R&D and the market, improvement of market operation and global operations.

With the growth of the global economy and pressures on natural resources, existing techniques are not enough to meet requirements for the sustainable development. Thus, such action plan advocated to absorb more private and public investment to be used for development and demonstration of the environmental technology and promotion of innovations, so as to bring inventions and creations in laboratories to the market.

EU adopted the following two major innovation actions: to set up the technical platform and the testing network, which is designed to know how to establish the private and public partnerships and how to make researches closer to the market. EU's existing member states and Norway have formulated roadmaps for implementation of the Environmental Technologies Action Plan. For example, in Spain's roadmap, climate change, urban sustainability, improvement of the production process, water, energy and transportation are regarded as priority areas in implementation of such roadmap.

2.2 New Zealand's Roadmap for the Research of Environmental Sciences

Ministry of Research, Science and Technology issued New Zealand's Roadmap for the Research of Environmental Sciences between May 2007 and 2017.Such roadmap involved three subjects requiring more attention and six environmental research areas.

Three subjects are as follows:

(1) Systematic understanding and integration. In-depth understanding of the environmental system requires effective combination of multiple disciplines. For example, areas requiring improvement of the system information include: interaction of the ground water and the surface water; impact of the fresh water resource along the coastline ; hierarchy understanding on the ecological system of fishery management; understanding on biophysics, social economy and healthy urban design.

(2) Technology transfer and absorption. To successfully complete technology transfer and absorption requires to pay more attention to forward-looking disciplinary and method-orientated researches and improving managerial innovations, so as to deepen scientific understanding and enhance communication technology (such as visualization).

(3) Application of information systems. Application of information systems includes data collection, data management, data acquisition and utilization of the new data-collection technology .Without improved modes for data management, integration and absorption of various disciplines are impossible.

Six environmental research areas are as follows: global environmental change, land/water/costal zone, urban design and disaster, bio-diversity, bio-safety and the oceanic system.

2.3 Korea's Second Comprehensive Plan for Development of the Environmental Technology (2008–2012)

On May 6, 2008, Korea Commission on Science and Technology issued the Second Comprehensive Plan for Development of the Environmental Technology (2008–2012) jointly formulated by 11 ministries and commissions including the Ministry of Education, Science and Technology, Ministry of the Environment. Such plan was published every five year. Such plan gave introduction to background and significance of its formulation, reviewed achievements made by First Comprehensive Plan for Development of the Environmental Technology (2003–2007), analyzed the status quo and prospect of development of Korean and international environmental technologies, and proposed Second Comprehensive Plan for Development of the Environmental Technology (2008–2012) on the basis of description of key issues and solutions on development of the environmental technology and analysis of SWOT[①] elements, its major contents are as follows:

1. Vision

To achieve technology-led Eco-Utopia, so as to make Korea become power of new green economy and to achieve Korea revival.

2. Three major objectives

(1) To carry out strategic technology development to ensure the state-of-the-art environmental technology in the world;

(2) To conduct selective and concentrated fostering of excellent environmental protection enterprises with international competitiveness;

(3) To perfect environmental sustainability index embodying the harmony between human and nature.

3. Four major core strategies

(1) To continue to expand the scale of investment for technological development of the environmental protection;

(2) To strengthen infrastructure construction for environmentally friendly technologies;

(3) To make selective priority investment on technological development

① SWOT represents strengths, weaknesses, opportunities and threats of strategic environmentally friendly technologies boasting international competitiveness;

(4) To strengthen policy support to promote development of the environmental protection industry.

2.4 A Roadmap for a Living Planet of the World Wildlife Fund

On September 29, 2008, WWF issued the latest report A Roadmap for a Living Planet. In this report, WWF presented a global protection framework. Under such framework, WWF proposed objectives for protecting of biological diversity and the global reduction of human footprint , determined key areas, species and footprint areas under priority protection and called on all walks of the life to protect biological diversity and achieve the earth's sustainable development.

1. Protection of biological diversity

2020 goal for protection of the biological diversity is as follows:

(1) To protect and effectively manage the biological diversity of the most outstanding natural areas on the earth;

(2) To restore and flourish important species with typical significance in ecological economy and culture;

(3) To keep human's global footprint below the level of 2000 and it on a downward trend, especially in such areas as energy/carbon, commodities (agricultural products, meat, fish and timber) and water resources.

2050 goal for protection of the biological diversity is as follows:

(1) To protect integrity of the most outstanding natural areas on the earth to create a safer sustainable future for all organisms.

(2) To keep human's footprint within the capacity to support lives on earth, fairly sharing natural resources on earth.

2. To reduce human's global footprint

2050 goal: to keep human's footprint within the capacity to support lives on earth, fairly sharing natural resources on earth.

2020 goal: to keep human's global footprint below the level of 2000 and it on a downward trend, especially in such areas as energy/carbon, commodities (agricultural products, meat, fish and timber) and water resources.

Brief Introduction to a Roadmap for a Living Planet

Consumption of resources by humanity has greatly exceeded the earth's bearing capacity, with human activities drastically changing the earth's climate. All these have imposed threats on the biological diversity and contributed to the possible collapsing of the life support system. Millions of people already felt such consequence. In the global arena, no matter in rich countries or poor countries, people are faced with uncertainty of food safety and water consumption and growing vulnerability in coping with natural disasters and diseases. If we continue to do so, things can only get worse. However, the only way for ensuring survival and maintaining prosperity is to protect biological diversity and reduce impact of the mankind on natural habitats. For this reason, WWF set forth a scientific framework for protecting biological diversity and reducing impact of the mankind on natural habitats.

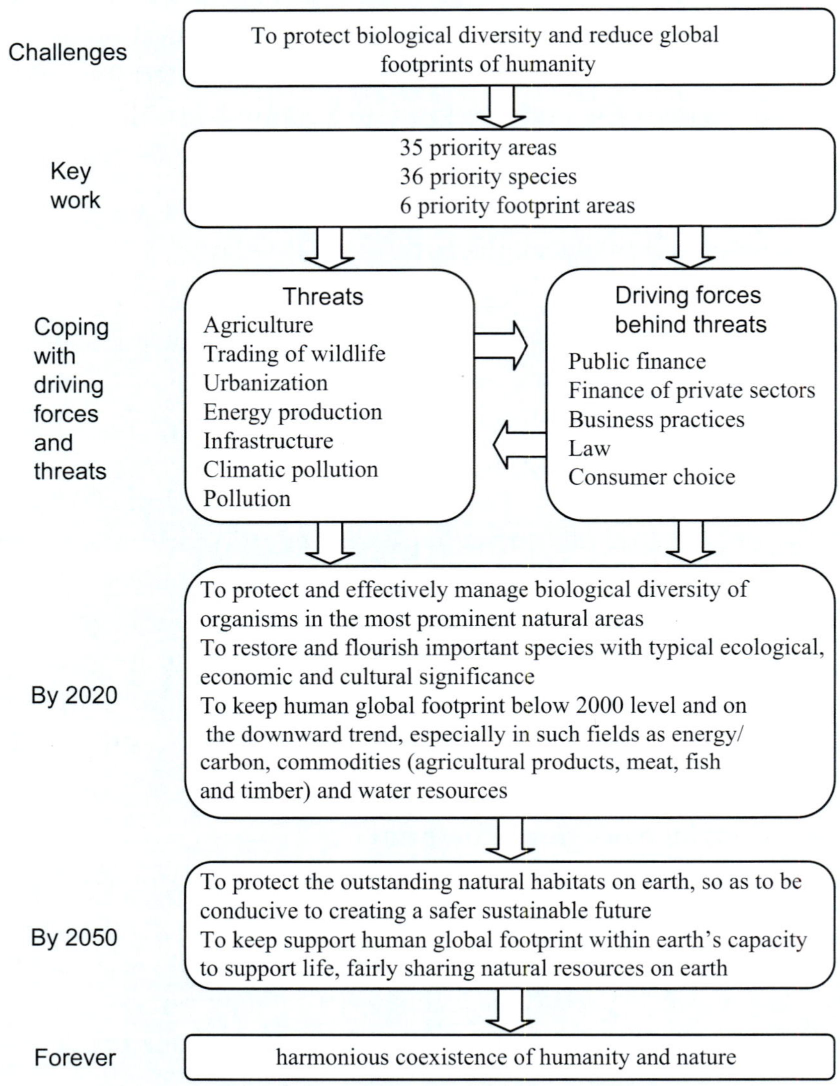

WWF global protection framework

Source of data: WWF. 2008. A Roadmap for a living Planet

2.5 Theme Action Plans-Phase I of the Natural Environment Research Council of UK

In December, 2007, NERC released the new strategy – Next Generation Science for Planet Earth: NERC Strategy 2007–2012[7].Its strategic goal is to carry out researches into world-leading frontier environmental science to enable society to timely respond to the global climatic change and the ever-increasing pressure on natural resources and to improve the leadership of UK in forecasting of regional and local environmental impact ranging from several days to several years, and to create and support research teams full of vigor.

1. Future Research Fields of NERC

NERC will fund multi-disciplinary researches at the international level to respond to key issues faced by the 21st century – sustainability of the human environmental system – including such environmental change problems as the climatic changing and loss of the biological diversity and sustainable utilization of natural resources.

NERC will also study on how to live with nature in harmony to reduce environmental changes, because that will be helpful for the long-term economic development of UK. For example, to improve environmental services and technological capability, develop economical, suitable and operable apparatus and equipments to monitor and control development and utilization of energy sources, thereby effectively avoiding losses due to natural disasters.

NERC will conduct more comprehensive cooperation with former partners, including cooperation with other research councils, Britain governmental departments and relevant international organizations.

2. Scientific Goal

Britain will take the lead to forecast impact of the environmental change on regions and places for several months or decades. In such scientific goal, its scientific challenge is embodies in broadening knowledge in observation, theory and environmental simulation. Such scientific goal mainly involves scientific research subjects, namely, the climatic system, biological diversity, sustainable utilization of natural resources, earth system science, forecasting and alleviation of natural disasters, environment, pollution and human health as well as monitoring and forecasting techniques. Focusing on such subjects, NERC will guide and fund the comprehensive multi-disciplinary research activities of several RCUK. To cope with challenges faced by social sustainable development due to pressures from natural resources and global climatic changes , NERC needs to work with multiple scientific partners traversing nature, society, economy and engineering as well as policy and corporate groups. Consequently, a new multi-disciplinary and multi-organization research plan, namely, Living with Environ mental Change, LWEC was introduced. LWEC is led by NERC and funded by ESRC and other research councils. In addition, such plan was also funded by Department for Environment, Food and Rural Affairs, DEFRA and other sponsors. The core of LWEC is to resolve needs of policy, enterprise and society for scientific evidence by combining natural, social and economic sciences.

LWEC also listed a series of middle and short-term priority research fields in science subjects attracting attention, and also determined middle and long-term priority research fields to be shaped, which are designed by continuing to intensify responsive funding for assisting and fostering the next generation of scientists.

At present, NERC has entrusted the theme responsible person to formulate theme action plans. Such subject action plans are designed to help NERC to define strategic researches to be funded, so as to cope with challenges brought about by the new strategy. In July, 2008, NERC management committee approved theme action plans of phase Ⅰ. Just like strategy itself, NERC will upgrade theme action plans every year. In the next months, NERC will implement new actions listed in the theme action plans, with the aid fund totaling 53 million pounds in the future years.

NERC will launch numerous large-scale research plans, including changeable water cycle, ocean acidification, alleviation of poverty and storm risks through the ecological system.

NERC will also formulate numerous small research programs, capability building and started plans. Such plans include: ① environmental nano-science research plan, which is designed to look at impact of nanoparticles in water, soil and air on the environment; ② fixed quantity research, which is designed to look at uncertainty of the climatic change and weather impact in forecasted regions and places; ③ carbon capture and storage, focusing on sustainability of CO_2 sealing; ④ understanding of results by using lands for renewable energy sources; ⑤ to work with Met Office to be responsible for numerous affairs of Joint Climate Research Program, so as to facilitate the sustainable development of such program.

NERC will also initiate multiple vision research programs to provide proposals for bigger investments in future theme action plans.

3. Expected Strategic Achievements

NERC will benefit from such objectives, that is because such researches will provide: ① recognition of new knowledge of environmental issues; ② studying and formulating the scientific basis suitable for the strategy to deal with environmental changes and avoid resultant losses ; ③ obtaining of profitable business opportunities.

Major scientific achievements include: ① climate forecast ranging from regions to areas and from seasons to years; ② trust levels about safe disposal of nuclear wastes and carbon capture and storage; ③ perfect location and the whole life cycle of renewable energy systems; ④ impact of environmental changes on services of ecological systems (such as clean water, air and food); ⑤ recognition of harms due to a sudden stop in thermohaline circulation; ⑥ reliable forecasting of storm intensity and its trace; ⑦ analysis of environmental factors for the spread of disease; ⑧ study on the role of biological diversity in determining functions of the whole ecological system; ⑨ improving of methods to monitor environmental changes and development of advanced technologies.

2.6 2008–2013 Scientific Strategies Formulated by Centre for Ecology and Hydrology, UK

1. Faced Scientific Challenges

Challenge 1: to provide the early warning for environmental changes through massive long-term monitoring and testing;

Challenge 2: to determine associative processes between changes of physical and chemical and ecological responses;

Challenge 3: to integrate the chemical process of ecology, hydrology, biology and earth and its feedback into models of climatic changes to improve the forecasting ability;

Challenge 4: to quantize impact of environmental changes on natural resources;

Challenge 5: to measure risks of exposure of ecological systems and humanity to biological, hydrologic and chemical threats;

Challenge 6: to formulate strategic and controlling means to alleviate the impact of environmental changes on ecological systems, ecological services and human health.

Environmental researches are mainly designed to achieve the sustainable development of economy. Thus, we shall formulate strategies and management polices to reduce impacts of environmental changes on ecological systems, ecological services and humanity.

Such strategies and actions should incorporate scientists, decision-makers and participators to ensure sound environmental management. CEH will formulate measures for protection, alleviation, adaptation and restoration, so as to protect ecological systems and human health from adverse impacts from climatic changes and other environmental risks.

2. Strategic Statement

Scientific strategies of CEH will be transmitted through scientific plans in such three interdependent science fields as biological diversity, water and biogeochemistry and supported by building of Environmental Information Data Centre, EIDC.

1) Biodiversity Plan

Biodiversity plan is aimed to determine changes and role of biodiversity in functions and resilience of ecological systems.

(1) To satisfy and supplement changes of the scientific theme of biodiversity of NERC, CEH will continue to provide the state capability building in massive long-term monitoring and investigation in terms of biodiversity. This will be achieved by recording and interpreting of diversity of genes, species and ecological systems and the changing in their distribution. In the whole process, technical advancement is required for improving the ability to detect implementation of new-type DNA methods and interpretation of satellite data.

(2) By closely working with EIDC, CEH will optimize data analysis and development through improving computing techniques and applying of

advanced information tools, and integrate new and existing biological and other environmental data to better quantize factors of physics, chemistry and anthropology in terms of biodiversity. All this will enable CEH to determine how such driving factors influenced the near real-time ecological evaluation of environmental and ecological systems.

(3) CEH will protect and restore practice in natural and artificial environments by continuing to adopt large-scale tests and supporting execution of management systems. Through such evaluation, CEH will develop models to predict local, national and international impacts of driving factors for individual or multiple environmental changes.

(4) Researches conducted by CEH will be conducive to better understanding of biological evolution and interactions behind large-scale biodiversity changes. CEH will also quantize the association between biodiversity of soil and that above ground, and promote legislation on EU Water Framework Directive and soil protection by determining impacts of such factors on the geochemical cycle and water flux, as well as conduct the overall evaluation of changing in soil utilization including degradation of habitats.

(5) By combing with other science plans and cooperating with EIDC, NERC will evaluate impacts of slow and rapid environmental disturbances on biodiversity. All this will provide better evaluation ecosystem services and detailed data for developing comprehensive evaluation of risks from land and fresh-water biodiversity.

(6) Based on powerful statistical analysis and modeling data, CEH will broaden knowledge on how to influence biodiversity and ecosystem functions by use of natural resources, and formulate reliable indicators to evaluate environmental conditions. In addition, CEH will be dedicated to maintaining its leading role in creating knowledge and provide advices on policy execution based on scientific basis, including Environment Stewardship Scheme, UK Biodiversity Action Plan and EU Habitats Directive.

2) Water Plan

(1) To achieve the sustainable utilization of water resources, CEH will comprehensively know the relationship between air, soil and water from different time and space scales. CEH will also continue to provide support for formulating and implementing of water polices (including EUWFD) at UK and EU levels.

(2) CEH will expand and improve professional knowledge on observation of the surface water body including appearance, biology and chemistry of rivers and lakes in lowlands and highlands. Researchers will use observation data to determine the trend in water-body environmental development and evaluate existing models or create new models. Additionally, CEH will set up the flagship station for outdoor observations, which is a part of cooperation between UK and EU networks, universities and other NERC research centers. All this will enable researchers to conduct tests on new technologies and measurement of association in the evolution process as well as make appropriate development

and utilization of scale models.

(3) CEH will carry out monitoring, test and simulation, so as to define the life and non-living processes and their interactive role in the fresh water body. By combing with CEH's biodiversity plan, researchers will evaluate consequences of biodiversity changes in the evolutive process of fresh-water ecosystems, and determine potential key species extremely sensitive to environmental changes. Then, they carry out new analysis such as data integration at the basin level via EIDC.

(4) CEH will use new in-situ sensor technology to provide high-frequency data. Researchers will, on a broader space scale, focus on depiction and quantification of fresh water and oceanic environments as well as the evolution process, interaction and feedback mechanism between land and fresh-water ecosystems. Therefore, CHE will enlarge its cooperation with British Geological Survey in measurement of interaction between ground and surface water bodies and the cooperation with Proudman Oceanographic Laboratory.

(5) At present, feedback between the water cycle, the earth's surface and the atmosphere and depiction on understanding of the evolution process in the global climatic model have some limitations. Consequently, CEH will resolve the uncertainty problem in forecasting of potential weather changes. Researchers will monitoring and testing data ranging from small regional scale to basin scale to reduce such uncertainty factors and test generation of models.

(6) CEH will enhance its simulating ability to correctly evaluate possible influence of the global change on availability of water resources. Specifically speaking, CEH will evaluate and quantize the microbial process, vegetation, soil and carbon storage and release in basins and the interrelationship among water influxes. Owing to profound impacts of flood and drought on the social economy, CEH will quantize such influences according to expected or predicted land utilization and climatic changes.

3. Biogeochemistry Plan

(1) CEH will continue to conduct monitoring activities including concentration measurement and depositional calculation of atmospheric pollutants. CEH will fabricate high-resolution maps of such pollutants and make the national budget, so as to meet requirements of Britain international reports such as National Emissions Ceilings Directive. CEH will monitor chemicals and wildlife in soil and water to set up a long-term development trend of time and space. In the process to monitor and analyze Britain pollutants, CEH will closely work with experts from National Centre for Atmo spheric Science and universities. Additionally, CEH will make positive contributions to science themes, controlling of environmental pollution and human health.

(2) CEH will carry out long-term tests including the big-scale field experiment of increasing of nitrogen and ozone in the semi-natural ecological system. Such experiments will enable researchers to quantize changes and channels of major nutrients and pollutants in the biogeochemical cycle and their impacts on animals and plants. Through EIDC, CEH will integrate data from

monitoring and testing activities to provide early pre-warning of environmental changes.

(3) In order to grasp conditions of long-term discharging and the settling influx of atmospheric pollutants, CEH will quantize chemical influx in the atmosphere and aerosol and improve knowledge on controlling of such process. Acquired recognition will applied to controlling of radioactive species, persistence organic pollutants and carbon enrichment and acidification.

(4) Knowledge related to the biogeochemical process will be developed and applied to the model development of the global climate (the earth system), so as to reduce uncertainty of the present weather forecast.

(5) Integration of monitoring and process research in models will provide necessary information for impact risks and services of ecosystems as well as human health. Researches conducted by CEH will be conducive to soil protection, because soli serves as a key interface for the biogeochemical cycle and changing. CEH will make development and on-site verification of quantitative chemical models for predicting behavior of pollutants and risks of their adverse impacts.

(6) CEH will quantize strategic responses of diversified land utilization to determine impacts of carbon sequestration and discharge. Land-utilization strategy will influence biodiversity, evaporation of soil moisture and amount of water in basins. Thus, for the purpose of realizing the sustainable land utilization, CEH will strive to provide comprehensive evaluations and policy suggestions.

4. EIDC

(1) CEH has collected rich environmental information in terms of environmental monitoring and studies made over the past decades. Boasting a large coverage range, such data is usually unique in the global, mainly including: bio-recording center (National Biodiversity Network Gateway included), UK Butterfly Monitoring Scheme, findings in the British countryside, Britain land cover map, NERC biological and environmental information center, phenology network, environmental changing network, observation plan of raptors and atmospheric data set.

(2) In cooperation with other NERC data centers, EIDC will not only improve utilization of NERC environmental data, but also being the major executor of EU Infrastructure for Spatial Information in Europe IN- SPIRE. Likely, EIDC is an important step in Shared Environmental Information System.

(3) EIDC will provide extensive data directory to facilitate data discovery. It will realize unified standards of data, if possible, and allow integration of different CEH data and data from other special geographical space portals. EIDC will also develop advanced delivery and visualization tools to convert environmental data into environmental information.

(4) Based in understanding of the processes of three major scientific plans, EIDC will quantize and demonstrate the mode, distribution, trend and change in the atmosphere, land and fresh-water environments. This will enable CEH to

develop tools for the near – real-time evaluation on environmental changes and their impacts. EIDC will also combine in-situ research and investigation and remote sensing data to provide accurate evaluation and forecasting of impacts of land utilization and climatic changes on ecosystems and ecosystem services.

(5) EIDC will integrate existing scattered data set such as chemical influx, and combine such data with other data set such as data about water and biodiversity. Potential key points within core NERC science programs is mining of cross data by use of existing data, and all this will enable researchers to evaluate impacts of climatic changes and responses of relevant ecosystems.

2.7 U.S. Long Term Ecological Research Network and National Strategy of the Ecological Observation System

1. U.S. Long Term Ecological Research Network (LTER)

Funded by National Sanitation Foundation (NSF) LTER has become the pillar of largescale ecological scientific researches, which none of people participating in building it 30 years ago has expected. Nowadays, with the emergency of new complementary networks such as NEON, GLEON, WATERS and OOI, potentials for the cooperative development of ecological researches are unimaginable.

Cooperation between physical networks and between disciplinary networks is likewise no less important. Development history of LTER proves that probing into other subjects especially social science and behavioral science is key to resolving present ecological problems, LTER network is developed with the growth of SBI, with obvious SBI brand in science of long-term ecological studies. LTER network has gradually covered researches into anthropogenic factors. Generally speaking, research groups especially the long-term ecological research groups have realized those important and remotest impacts which are sometimes behind human activities.

LTER network has included innovative challenges with development prospects which are highly relate to internetworking age: motion acknowledgement of social and environmental sciences and attempting to the socio-ecological relationship between organisms, material cycling processes and ecosystems which cross various space scales. Ecosystems serve the human society, usually, humanity can clearly note such service, however, sometimes significance of ecosystems may be underestimated or even completed neglected by mankind. How to correctly understand importance of ecosystems, how to guide and then change our actions by using such viewpoints and how to influence appearances and functions of ecosystems are of vital importance to our understanding of ecosystems that humanity is dependent for survival and to their sustainable development. Ecologists, geophysicists, sociologists and environmentalists shall carry out interrelated, harmonious and interdependent studies to completely know various relations between ecosystems and the human society.

Type and intensity of connection between nodes of the test network and sites required for interdependence are the important challenge faced by building of associated disciplinary studies. For this reason, LTER is actively searching interrelated cooperative websites, and if possible, will realize sharing in network infrastructure and other public resources, thereby facilitating collection and acquisition of environmental data. Large-scale connectivity scientific research is based on local scientific large-scale coverage research, otherwise, scientific hypothesis related to connectivity of socioecology can't be precisely examined. Many cases in the early of research of socioecology (many descriptions of the present period) show that new networks in fields of excellence environmental observation and connectivity science will be able to fully define the key connection of global scattered ecosystems.

All these websites have realized the importance to integrate environmental information from various websites, only by this can we resolve the most stern ecological and environmental problems, namely, how to meet needs for the ever-close connected global sustainable development.

2. Continental Strategy of NEON

NEON is designed to resolve the following two problems:

(1) How do ecosystems with space scales (US) and their components respond to impacts of natural and human factors such as weather, land utilization and species invasion? What are response speed and mode?

(2) What kind of role do biogeochemistry, biodiversity, hydroecology, organic texture and function in researches of climatic change, land utilization and specifies invasion? How do these feedbacks change with ecological backgrounds and space scales?

NEON will provide data and other devices to support ecological forecasting at the continental scale to resolve foresaid problems. Required space scope ranges from genome to continental scale, with time horizon from several seconds to decades. Comprehensive behavior of countless organisms will control transmission and chemical characteristics of the atmosphere, regulate materials on the earth surface, and influence water supply and quality. Difference between the organic scale and its scale acting on the global environment triggered major problems in large-scale ecological researches. Lives in the natural environment are fruits of the longest time and the largest space scale, thus, we shall know the biosphere process based on quantity of organisms, ecological groups, ecological communities and creatures and recording of other scale phenomena.NEON is ready to resolve the problem of scale diversity through researches into processes, interactions and responses as well as connectivity. Most environmental monitoring websites only focus in processed and results, but not connect interactions and feedback actions. NEON focuses on the multiple-scale nature of biosphere. Starting with organisms, species groups and organic communities, Basic NEON monitoring begins to directly observe their biological processes. Limited funds restricted quantity and space scale of basic observation. For this reason, NEON adopted a cost-effective continental scale strategy to arrange observation units. Observation units must objectively represent designing of environmental changes to make systematic screening. The early US space maps divided the state into different ecological zones, compared to such early maps, zone-diving of NEON is based in new systematic analysis and application of database of state ecological and climatic changes. Moreover, the statistical design is based on variables of multiple geographical groups, the best result from analysis of geographical groups is to set up 30 subareas.

NEON data system will integrate local and regional measurement data to quantize processes and responses of climatic changes, land utilization and varieties of species. By use of integrated data and high-altitude remote sensing data, observatory scale for NEON website observation can be extended from regional ecological process and response to reflection of airborne telesensing and other geographical data.

NEON observation strategy provided controversial observation sites of physics and biology with strategic significance, through statistics, observation and design, such sites spread throughout the United States. In a word, such sites are set up to serve as the only effective means for sampling throughout the whole America. Such effective means will not only can determine overall changes of the state (through sampling, scale determination and observation design), but also serve as a telescope to observe nation's key internal key problems, differentiate diving mechanisms of different regional ecological changes or different responses to changes as well as sample vectors of matters, organic matters and energy transfer. By acknowledging that biology is both global and highly localized, NEON strategically supplemented the scale of the present observation system and coordinated life observation requirements in these two aspects. Although NEON design still fails to completely resolve all problems, as a research platform, NEON will become the liaison for observation, understanding and forecasting of environmental changes in the world of humanity.

2.8 Perspective Plan for Researches Conducted by NOAA in the Future 20 Years

1. Long-range Goal of NOAA

The goal is to build a information-based society, fully understand the role of oceans, coastal belts and biospheres in the global ecosystem, so as to facilitate formulation of best decisions for social and economic development.

Perspective plan for researches conducted by NOAA in the future 20 years provides the foundation for NOAA long-term research methods, and will ensure that NOAA will meet changeable requirements for society to information and products in the next 20 years and make the optimum decision making for social and economic development.

2. 2025 World – NOAA's Prospect in the Information Decision-making Field

By 2025, impacts of NOAA environmental forecasting on decision makers will generally exist, and vary in distribution of time and scale from hours to months. Benefiting from tremendous progress in forecasting ability, tornado and extreme weather events will almost no longer be harmful to the humanity, and nearly everyone will possess PEA capable of issuing early warning of various environmental hazards. By relying on PEA, captain will easily obtain the latest situation about the weather and ocean, and based on daily analysis of underwater marine fronts and whirlpools to determine the optimum fishing locations, thereby avoid sailors exposed to the dangerous environment.

Prior to 2025, the world will make optimum social and economic decisions by depending on reliable detained environmental information and forecasting. Such decisions will facilitate economic growth, increase amusement and recreation opportunities, protect the public health, significantly reduce rainstorms and alleviate damages caused by drought and long-term climatic changes.

3. Social Needs of the 21st Century and NOAA

In the 21st century, degree and urgency of social needs for information about atmospheric and oceanic conditions, conditions of existence of marine biological resources, water rights, marine navigation and weather forecast as well as comprehensive management means will sharply rise. Population growth in the global will bring about many new needs and influence economy, human welfare and environment. Impacts from globalization process and cooperative trend will cause increasing of social needs.

–Growth of American and global population will aggravate impact of the severe weather on human health, water rights, security and economic investments, with dependency and expectation of people on weather forecast increasing accordingly. The ever-complicated land planning will raise more requirements for NOAA data acquisition and analysis.

–In most places of the world, people's needs for protein will make importance of fishery resources continuously increase, thereby causing finance to further slant to management of fishery resources. Needs for understanding and forecasting of fishery production capability will increase, thus, we shall expand sea farming to meet people's needs for food protein. However, this will inevitably influence the economic and oceanic ecosystems. Threats on oceans and human health also include communicable diseases, for which we should strengthen monitoring of spread of medical wastes and modes of transmission of patients with infectious diseases. To seek marine biologics to cure non-communicable diseases such as cancer, multiple scleroderma and degenerative brain disorder, people will mine marine natural resources, this will also pose a threat on the marine ecological system.

–Human work, transport and recreation will continuously aggravate pollution, nitrogen and other by-products will continue to flow from lands to oceans, peak reproduction of harmful marine algae (such as the occurrence of red tide) will restrict utilization of marine resources. Thus, it becomes more necessary to supervise and regulate eutrophication from lands. Aggravation of species invasion caused by exchanging of water bodies and recreational fishing will increase biologic contaminations. We shall intensify monitoring, simulating and tracking of atmospheric pollutants from American and surrounding coasts.

–Even if industrialized countries such as America and Western Europe countries reduced emissions of greenhouse gases, emissions of carbonaceous gases and other greenhouse gases will continue to increase. Because greenhouse gases further aggravated the global warming, America and other developed countries will pay more attention to the atmospheric monitoring, and base on it to provide economic aids to reduction of emissions of greenhouse gases.

–Potential risks of abrupt accidents and discharging of biological, chemical and radiative toxins into the environment has been on the increase, this requires us to intensify observation and analysis and include observatory and analyzing results into the emergency response system of managers of federal, state and local emergency affairs.

–The ever-increasing international trade requires more bigger transport ships and infrastructures such as wider and deeper ship channels. Through information forecasting of weather, wind and ocean current, large ships can keep sailing along the optimum route, and sailing can be more quick and safer. However, local congestion caused by increased bigger ships will exert adverse influence on economy and society and arouse people's attention to the public security. To cope with such increased risks, we need to obtain positioning information, including real-time information such as position of ship, sea conditions, weather and oceanic cartography, so as to properly arrange arrival time of ships.

–In terms of agriculture, manufacturing and plant's location, economic importance of the long-term weather forecasting will continue to increase.

–Intensifying of understanding on ecosystems can meet social needs and make the government conduct efficient decision-making and responses. Thus, NOAA and its partners, academic institutions, private organizations and international partners need to understand and upgrade its structures and functions, so as to secure the sustainable development of oceanic and other natural ecosystems.

4. Techniques of the 21st Century and NOAA

With the social development, dependency on analyzing and forecasting of comprehensive ecosystems will continue to increase, and NOAA will continuously introduce new technologies to meet social needs. NOAA relies on four key techniques to conduct environment description, understanding and forecasting, such four techniques are sensor, platform, IT and telecommunications respectively.

–Sensors capable of acquiring biological, chemical and physical information have existed for a long time. However, many of such sensors are costly, less durable and can't be applied to the environmental monitoring. In the future 20 years, sensors will be improved in reducing of cost and increasing of durability, and a large number of degradative biosensors will be applied to monitoring of the target environment at a fairly low price. As for NOAA, in the next 20 years, techniques capable of being applied to rapid-changing conditions and playing dynamic role will be essential.

–Except for equipping the fixed platform with automatic sensors, NOAA will also develop mobile sensor platforms (such as robot airplanes and automatic submarines) to monitor and observe lands, air, sea surfaces, deep oceans and seabed. Such technology is especially favorable for submarine detecting and drawing. Platforms featuring advanced performance, complicated structure and the above-mentioned sensors will provide researchers and operators with techniques for complicated sampling. The pre-programmed intelligent system will make observation and measurement in a highly optimized manner.

–With the operating speed of computers doubling every 18 months, IT will continue to develop, expecting to create complicated simulating systems and tools for data management and analysis. NOAA will develop simulation-based analysis techniques through data assimilation and analyze data obtained by new-type sensors. NOAA will adopt complete high-resolution modes, including information about land foundations, coasts, marine biological resources and the atmosphere. Through such complete modes, NOAA will be able to depict, understand and predict interactions among various components in a higher resolution.

–Telecommunication technology will continue to improve its resolution, bandwidth capacity and effectiveness. As the key to facilitating popularization and application of sensors, resolution of the GPS of telecommunication technology will reach 1cm in the future five years. The global network will easily realize seamless connection of modeling and ecosystematic information center, service providers and users. Popularization of PEA and powerful data communication system will enable users to reach back data to high-performance computers, and use modern simulating and forecasting techniques to meet their individual requirements. Society will become more and more dependent on the telecommunication technology, NOAA space weather forecast is of same importance to managers and users of telecommunications.

The above-mentioned all techniques will be applied to developing of a complete GEOSS. The existing observatory systems are composed of independent parts based on needs for various environmental information. Many of such observatory systems are set up for one single purpose, and connected via various networks including data formats and assimilation methods. Integrated observation systems will be able to analyze and predict conditions of the atmosphere, lands, rivers and oceans, with the focus on elements such as hydrology, biology, geography and chemical cycles constituting ecosystems.

5. Product and Service Examples of 2025 NOAA

Owing to great progress in integration of information, product and service, NOAA will provide the American public with easy-to-use, seamless and complete information products and services, which will inevitable bring out an unimaginable reform in daily lifestyles of the American public: forecasting of extreme weather events and air quality will be greatly improved; reliability of long-term weather forecast will continue to increase; management of fishery, harbors and resources protection will be improved based on scientific information; various vehicles of sea, land and air will be able to achieve real-time obtaining of important environmental information. Major progresses in these aspects are as follows:

–Pre-warning of strong storms and weather events will save more lives and properties. The future advanced information transmission systems will be able to transmit pre-warning information about storms and weather events, with the pre-warning system significantly improved. For example, pre-warning of frequency of tornado will be hour-level, but not minute-level. Techniques such phased-array radar, improved understanding on the meso-scale weather process and mode development are foundations achieved by us. Improvement of ability of emergency pre-warning of storms and monitoring of tsunami will greatly reduce damages caused by such disasters on life and property.It is worth mentioning that through developing and improving mode-based flood-submerged drawing, develop the deep-ocean pre-warning system, ability of NOAA to protect the public from tsunami disaster will improve in a certain extent.

–Air quality forecasting will be upgraded and extended throughout the country. Air quality models will include not only pollutants such as ozone, but also all chemicals and fine particles related to "chemical weather". Such models will cover the country's territory, and be used to resolve local and regional problems related to the human health and air visibility. Future air quality models will be used to assimilate observation results of various chemical constituents, and improve the forecasting system for preventing the public from damages due to point source diffusion pollutants and extend air quality forecast to several days and even a longer time.

–Improvement of air quality forecast will not only prolong time scale of forecasting of storms, tsunami and floods, but also create substantial economic benefits for the American society. Application of more timely and accurate forecasting, pre-warning and information will save losses of life and property totaling 1 billion dollars.

–NOAA will conduct weather forecast ranging from seasonal scale to age scale, and explicitly define its uncertainty level. To do this, we need to increase observation data, improve understanding on the earth system, speed up popularization of IT and telecommunication technology and reduce their costs.

–With the explicitly defining of uncertainty, products related to weather will gradually turn to the regional level. NOAA will launch and improve researches into understanding of key process, to reduce uncertainty of forecasting and improve knowledge about forecasting ability and limitations. Statement of climate results will become the standard, and the architect for usage of weather products by decider and the pubic. Paying more attention to application of seasonal climate forecast will be helpful for resources managers to alleviate impacts of flood and drought on agriculture and other aspects and better control water resources, thereby preventing spread of diseases extremely susceptible to temperature.

–The annual worth of American industries susceptible to weather and climate nearly totals 3000 billion dollars, involving finance, insurance, real estate, retail trade, wholesale trade and manufacturing. Other water-consumed industries or non-water-consumed industries such as agriculture, energy, water supply and public health, flood control and navigation have become the pillar of the American economy, however, so far, many large reservoirs of America have failed to develop due role, that is partly because of exact forecasting of seasonal climate and runoff still fails to be achieved.

–NOAA will provide the scientific support for using ecosystem theories to manage coastal and marine resources, so as to utilize comprehensive and reliable science information to resolve complicated social problems. Vision planning of NOAA includes: exploring of various important ecosystem directives; seeking of highly automatic observation systems measuring such indicative factors; building of inter-type balanced models for evaluating stress origins of various ecosystems and social costs and benefits. In a word, we want to transform research findings into scientific knowledge to enrich highly technologized and computerized decision support tools, so as to provide significant, sensitive and reliable scientific suggestions and information products and further use ecosystem approaches for effective management.

–Studies on ecosystems will be beneficial to improving knowledge about chemical, physical and biological interactions, and enhancing the ability of ecosystems and environmental change models, all this is of significant importance to reducing uncertainty of structures and functions of ecosystems. Observation ability will be upgraded to develop physical-biological coupling model at the multi-space scale. Understanding of the trophic level and biological chain will be improved and finally linked up with climatic changes. Based on such information, NOAA will make effective forecasting of a complete set of ecological conditions, including fisheries, anoxia, harmful algal blooms, coastal degradation and water quality problem.

–Building of better costal belts and marine decision support tools and probing into abundant marine resources will facilitate the shift to ecosystem-based management mode. The key point is to better manage costal zone resources, reduce anthropogenic influence, ensure the sustainable development ability, enhance human health and improve the quality of life. Improved scientific information will help people to effectively deal with various problems such as production of marine food, harmful algal blooms, coral degradation and worsening of ecosystems.

–NOAA will lead the marine detection plan beneficial for the state. For humans, there are many unknown mysteries in oceans. Marine detection will be characterized by unknown fields of physical, chemical, biological and geographical, and designed to provide new hypothesis for ecosystem and climate researches. Discovery of shipwreck relics and other cultural resources sinking into the sew will help to know the past of modern and to explain unknown mysteries. Human will enjoy unexpected opportunities to benefit from oceans, oceanic creatures may be the medicine treasure, with new energy and food resources explored. Marine exploration will be helpful to testing and developing of new-type sensors and platform technologies, as well as improving human's understanding on the marine process and resources.

–Any land or sea or air vehicle will obtain real-time weather, positioning, terrain and other relevant environmental information. Information and telecommunication technologies include the progress of high-resolution GPS, which will make foresaid imaginations come into true. In America, traffic accidents caused by poor weather conditions amount to 1.5 million, with 800,000 people injured and 7000 deaths. Additionally, serious delay of personnel and goods during transportation also caused huge economic losses. About 70% of airport delays were caused by adverse weather, with the annual loss totaling 4.2 billion dollars, however, if with effective observatory means and weather forecast, such personal casualties and property losses may be avoided. Improved traffic, weather forecast and navigation will greatly curb occurrence of accidents and reduce economic losses. In order to meet traffic requirements, NOAA's researches will be designed to integrate real-time observation, marine on-site report and forecast, and utilize advanced visual, electronic and automatic decision support tools and services for location tracking information. Such researches will focus on improving skills, applicability, evaluation on uncertainty, positioning precision and visualization means of models, especially on development of new-type electronic nautical charts.

Improving of navigation tools, real-time observation, on-site report and forecasting of marine products together with GPS-aided navigation will provide safer and more efficient harbors for the society, and enlarge harbor capacity without reducing passenger safe space and harbor infrastructure and without destroying animal habitats and ecosystems.

6. Status of NOAA Researches

In order to better understand NOAA research planning, it becomes necessary to analyze the research status of NOAA from the perspective of structures of formulated objectives and targets, including ecosystems, weather, climate, commerce and transport (as shown in Table 2.1).

Table 2.1 2025 NOAA product and service cases

Ecosystems	Related to forecasting and alleviating strategies: anoxia, harmful algal blooms, beach closure, invaded species, sea waves, quality and quantity of air and water Ecological evaluation and forecast related to climate (such as coral decolorization) Decision support tools for proper fishery management, shoals development and management of marine resources at the ecosystem level Evaluation of the impact of the sea level changing on coastal zone resources and ecosystems Better incorporation of observation data used by managers of coastal zone ecosystems Forecasting of production capability of fishery to be influenced by the climatic changing

(Continued)

Climate	Improved weather forecast based on models of the earth system and ranging from intra-seasonal, seasonal to yearly scale Forecasting of water resources and drought by considering loss of nutrients Forecasting of diseases related to weather (such as malaria, SARS, West Nile virus) Forecasting of the sea level changing Programs for alleviation and adaptation of the future weather, including the changing of land utilization Decision support system for climate information and evaluation
Weather and water resources	Providing the weather forecast with partners at the neighbor level, and forecasting of 10-14 days which are as accurate as that of the previous 7-10 days Forecasting of tack of big storm and tornado of the county level Modes of forecasting of air quality and chemical elements at the regional or continental scale Improved forecasting model for river flow, forecasting of the flow level from drought to flood, and its interactions with underground water, utilization of water resources, harbors and coasts New-type soil moisture forecasting model for pre-warning of agriculture and the debris flow New-type systems for protecting the public from hazards of special point source discharge
Commerce and transport	Building of real-time atmospheric delay model (troposphere and ionosphere), so as to enable precision of real-time GPS positioning to reach the centimeter level Utilization of high space-resolution and accuracy measurement data to support researches into safe navigation and ecosystems Developing of vehicles which are capable of detecting and responding to environmental changes and tools for obtaining real-time weather information and formulating of routes Strengthening of new apparatus, new technology and new process of security and capacity of the aviation management Manufacturing of tools beneficial for transport time, reliability, efficiency, goods transportation costs and the environmental health.

7. New Model for the Global Ecosystem Forecast

Looking into 2025, NOAA products and services will be significantly improved on the basis of present researches. Following activities will facilitate the goal realization:

–To monitor and observe lands, oceans, the atmosphere and space, and set up the observation and data acquisition network for tracking changes of the earth system;

–To understand and depict how the natural system works through information survey and interpretation;

–To evaluate and forecasting the changing of such systems and provide information about the future;

–To absorb more individuals, partners, social groups and enterprises and make them to become information users of NOAA, as well as provide assistance in their utilization of NOAA information;

–To manage costal zones and marine resources, so as to benefit the environmental friendly economy and public security to the greatest extent.

3 China's Socio-economic Development Trend and the Needs for Ecological and Environmental Science & Technology in the Next 50 Years

3.1 Population Growth Trend and Its Driving Force

The population growth in China has significant stage characteristics. Particularly after 1980, China's population growth rate gradually slowed and had an unprecedented slowing down after 2000. Therefore, considering stage characteristics and population growth feature, logarithmic model is not an ideal model for prediction. According to the research on population prediction model, the growth model has a better effect.

Apply curve fitting analysis tool of SPSS software in Logistic curve fitting of the population data from 1949 to 2007. In accordance with the prediction of United Nations Economic and Social Affairs Population Division, China's population will reach the peak in 2030, that is, over 1.45 billion. Following the trend to 2050, the population will be around 1.5 billion.

According to such growth trend, in 2020, the population will be controlled within 1.45 billion; it will exceed 1.45 billion in 2030 and reach 1.5 billion in 2050. Population growth gives rise to a series of socio-economic and environmental problems such as employment, material supply, social security, energy consumption and increasing pollutants discharge, etc. Human impact on global biosphere controls a number of main aspects of ecosystem function. Currently, most of the world available fresh water, arable land, fishery output, nitrogen emissions, carbon dioxide balance and reproduction are under the control of mankind. Human activities have greatly affected the outcome of evolution and significantly speeded up the changes of species evolution at the same time.

3.2 Urbanization and Its Driving Force

It can be seen from the total urban population change from 1949 to 2007 (Fig.

3.1) that the growth of urban population shows a strong regularity, and the Logistic model can be used to determine the upper limit of Chinese urban population so as to predict the total urban population. Based on the existing research, in 1980s, when China's urbanization rate reached 25%, it actually entered the medium rapid developing phase of urbanization. The urbanization rate will reach 55 % in 2020, 65% in 2030 and over 70% in 2050. After reaching 50%–60%, China's urbanization rate will enter the slowing down period and tend to be stable. It is foreseeable that in the next 50 years, China will continue the rapid development of urbanization and gradually slow down in the medium and late period.

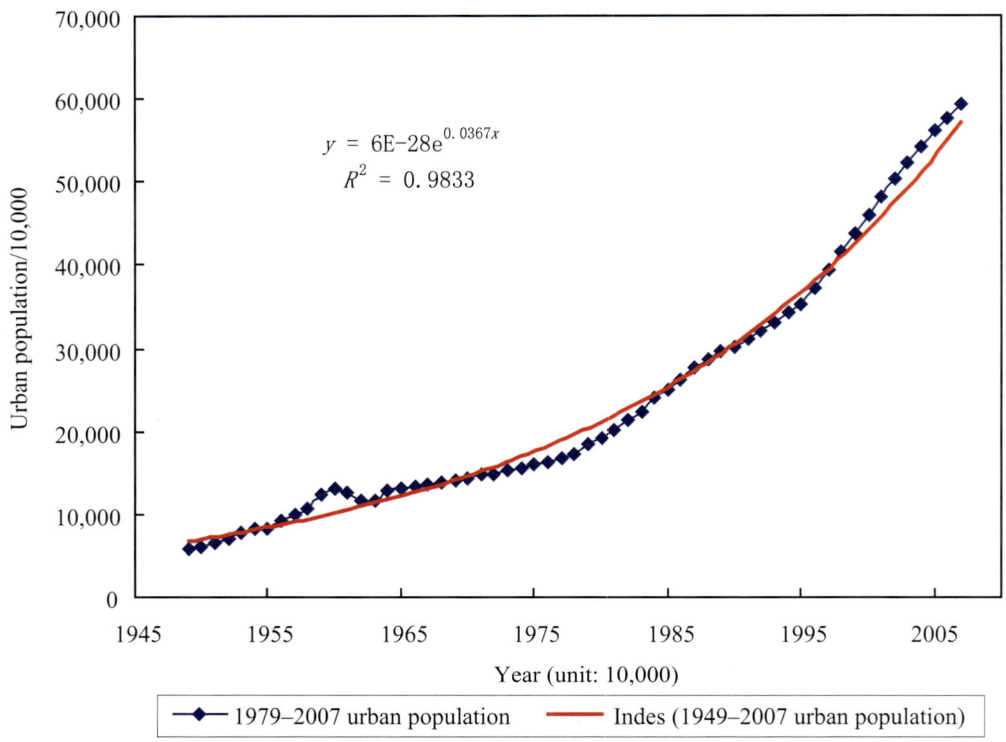

Fig. 3.1 The change of Chinese urban population from 1949 to 2007

Urbanization has brought increasing intensity of per unit area metabolism and energy consumption, changes in land utilization as well as spreading infectious disease due to concentrated population. In China, the annual increase of urban land use is 150,000 hm^2, water consumption increases 2 to 3 million tons, energy consumption increases 1.75 million tce, domestic wastewater discharge increases 1.8 million to 2.4 million tons, waste gas pollution will also increase significantly, and solid waste will increase at the rate of 2.6% annually, etc. Therefore, the city's ecological environment is facing tremendous pressure during the urbanization process. If these issues can not be well resolved, the impact will continue and the sustainable development will be restrained consequently.

National urbanization trend is in the fast developing stage before 2020, which will boost the development of urbanization and environmental quality technology. Moreover, according to the regional difference in national urbanization, carry out research on the four major urban agglomerations along the eastern coast (Zhujiang River Delta city group, Yangtze River Delta city group, western strait city group, and group of cities around Bohai) before 2020; from 2020 to 2030, focus on the study of central city groups (Changsha, Zhuzhou and Xiangtan City Group, central China city group and Wuhan urban agglomeration); then gradually extend the research scope to western area from 2030 to 2050.

3.3 Economic Development

1. Per Capita GDP Growth

GDP per capita is a primary indicator for measuring the level of economic development of a country. Seen from the index of per capita GDP growth trend of China from 1952 to 2006 (100 in 1952) [Fig. 3.2 (a)], China's economic development has generally gone through three phases: the first phase is from the founding of New China to the beginning of reform and opening up, the actual per capita GDP growth is smooth and slow; the second phase is from the reform and opening-up to 1990, the actual per capita GDP growth starts to leap forward; and the third phase starts from 1990, the actual per capita GDP grows at a extremely rapid rate. There is a large fluctuation in the first phase [Fig. 3.2 (b)], while the growth in second and third phases is quite stable, which can be accurately fitted by the index model respectively [Fig. 3.2 (c) and Fig. 3.2 (d)].

China will reach the level of medium-developed countries around 2050. An index growth model can be set up based on the data since reform and opening-up in 1978. Seen from Fig. 3.2(e), the model can accurately fit the situation of China's economic development in the past 30 years, thus the economic development trend in next 30 years can be predicted.

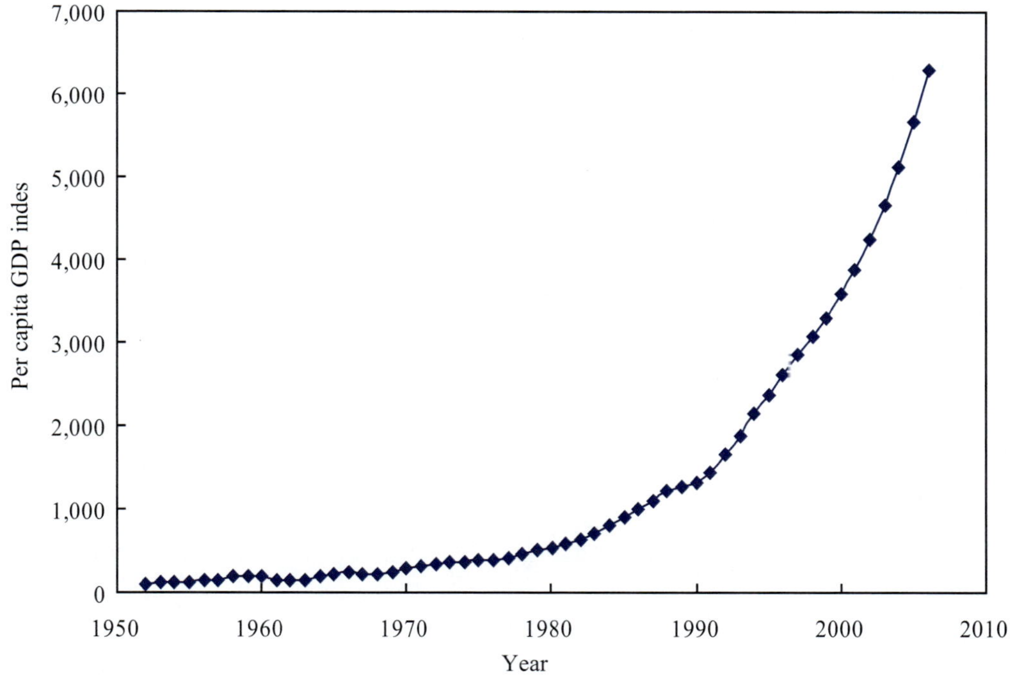

(a) 1952–2006 per capita GDP index

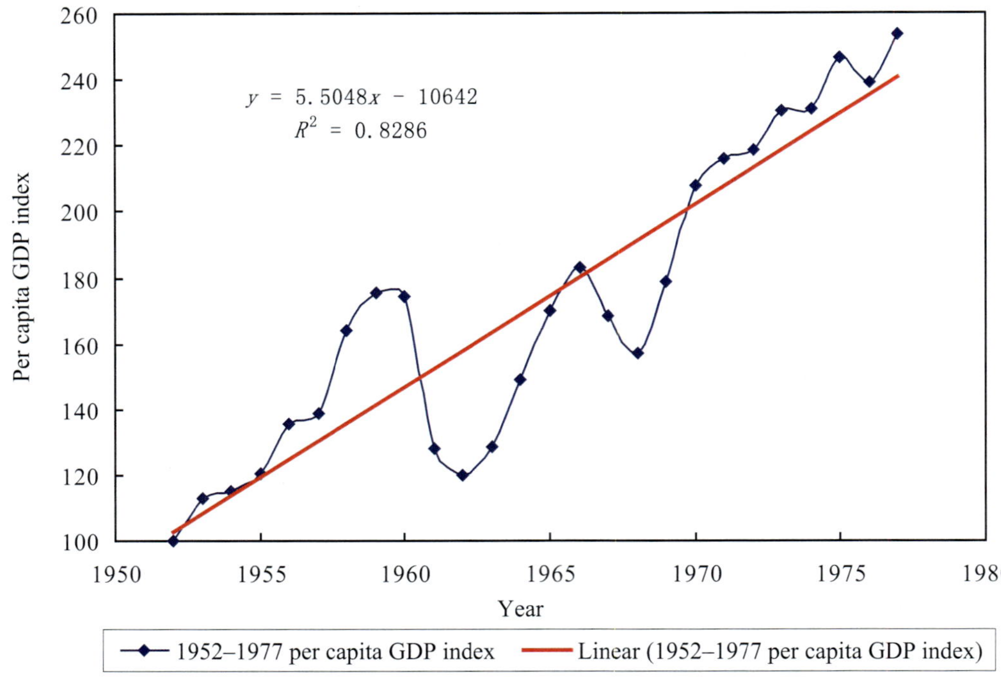

(b) 1952–1977 per capita GDP index and the fitting

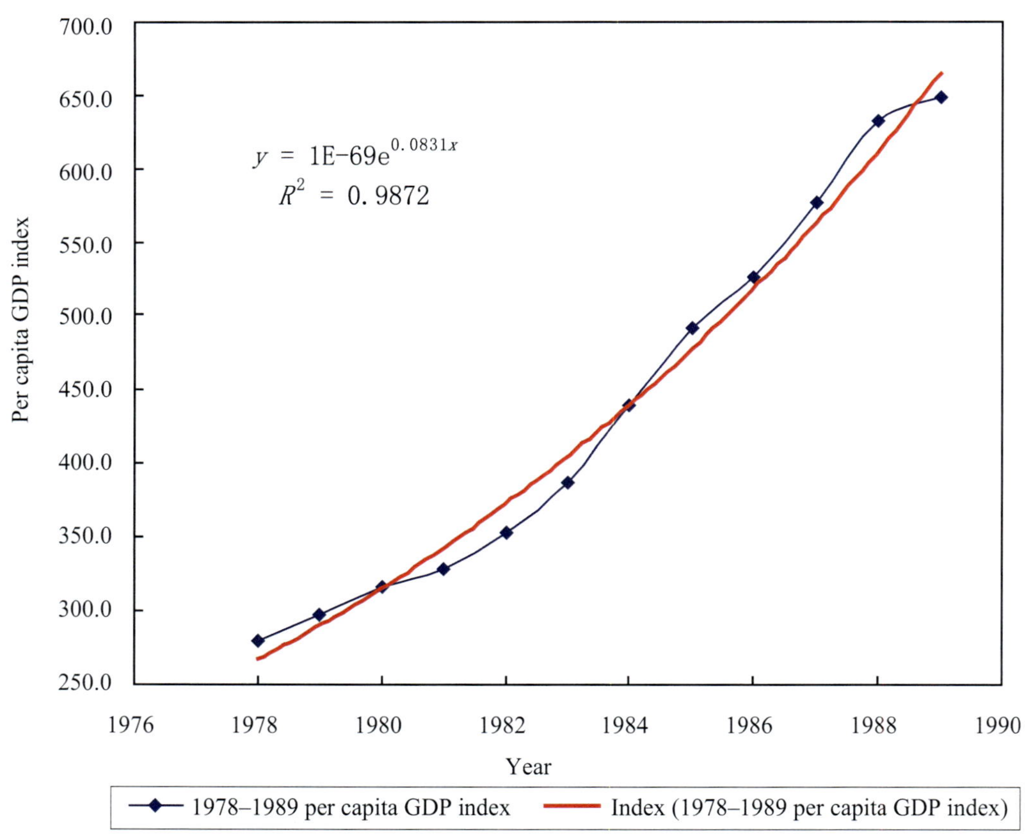

(c) 1978–1989 per capita GDP index and the fitting

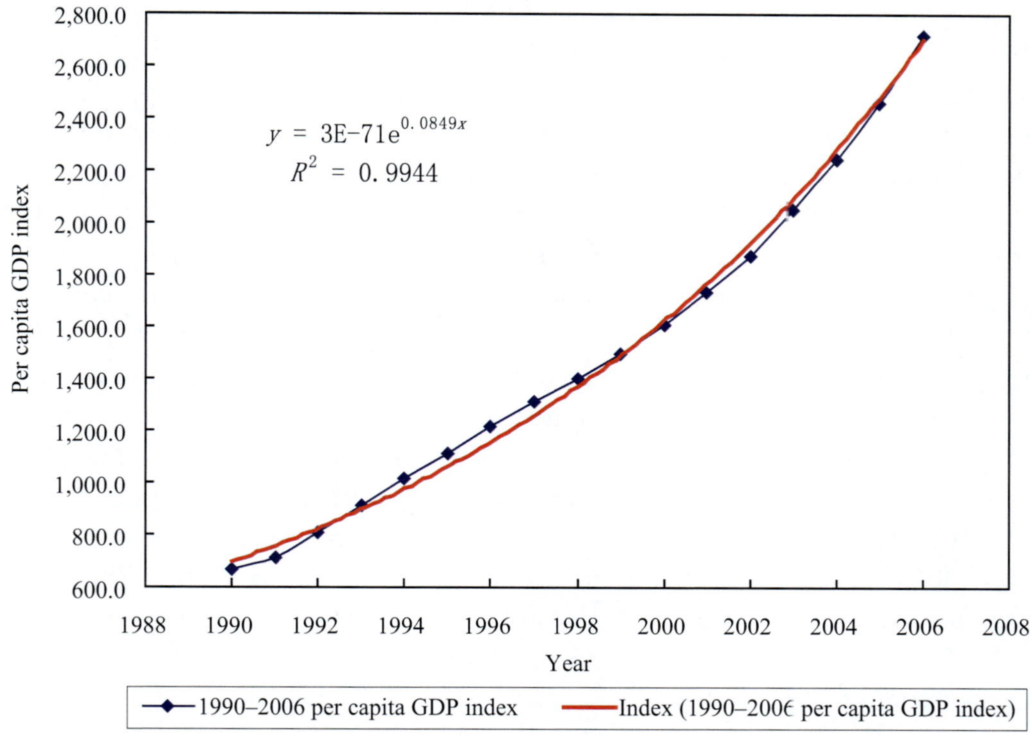

(d) 1990–2006 per capita GDP index and the fitting

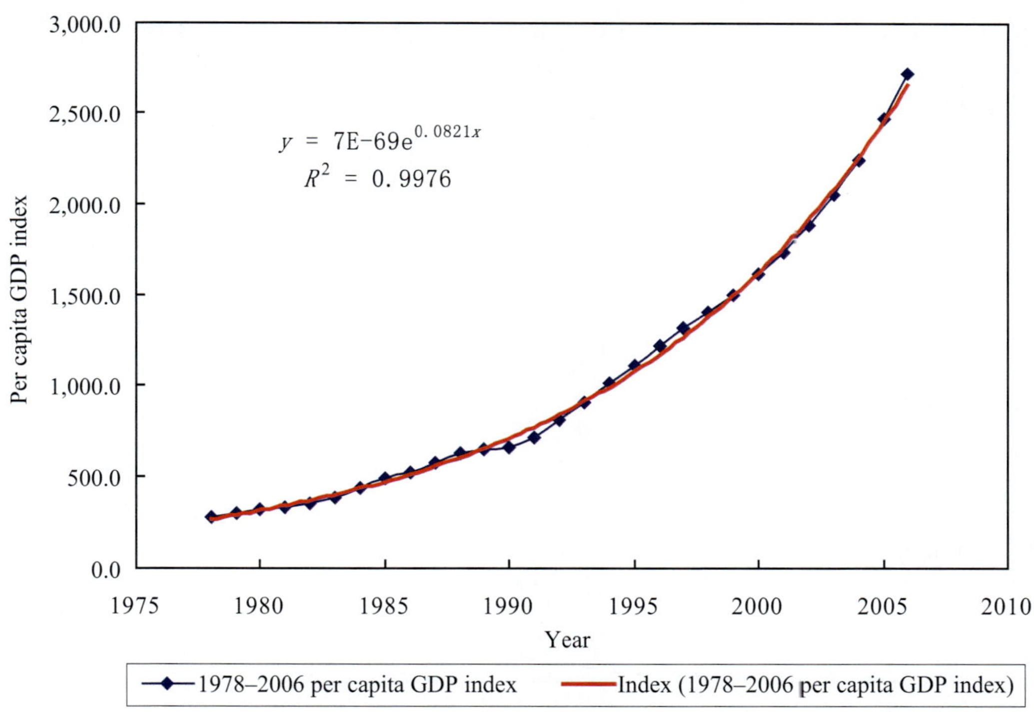

(e) 1978–2006 per capita GDP index and the fitting

Fig. 3.2 1952–2006 per capita GDP index and the fitting in China
Source: Fifty years of new China, the compilation of Statistics (1949–1999), China statistical yearbook, 2007

The model may overestimate the level of economic development. Moreover, it is impossible for the actual per capita GDP to enjoy infinite growth, for it will reach the limit at a certain point of time. Hence, the approach of two-phase multi-model prediction is applicable to forecasting the economic development in next 50 years.

In the first phase, from now to 2020, employ index growth model [Fig. 3.2 (e), $y = 7E\text{-}69e^{0.0821x}$]. According to the prediction of this model, the per capita GDP of China in 2020 will be 4.6 times of that in 2000, namely, 34,891yuan.

In the second phase, from 2020 to 2050, employ Logistic growth model for prediction and the key lies in setting the upper limit. Draw on the current per capita GDP of the developed countries (in 2005, China's per capita GDP was only $1,700. While according to the rank of World Bank, the per capita GDP of Luxembourg, ranking No.1, was $43,940, and that of America and Japan was $37,610 and $34,510 respectively). Considering the actual situation in China, take the per capita GDP of America as the upper limit for reference. That is to say, the upper limit is about 22 times of that in 2004, so the ceiling is 49,339 USD, 22 times of 2,242.7USD in 2004. Therefore, 50,000 USD can be regarded as the upper limit.

According to this upper limit, the model y = 1 / (1 / 50 000 +1.99 E +49 * 0.941406 ^ x) can be obtained through the fitting of SPSS software. Meanwhile, in order to be cohesive with the first phase, the value difference of the two models in 2020 is 3,106.862, which should be added to the predicted value of the second phrase, thus the per capita GDP in 2050 will attain 12.8 times of that in 2000. The whole prediction model is shown in Fig. 3.3.

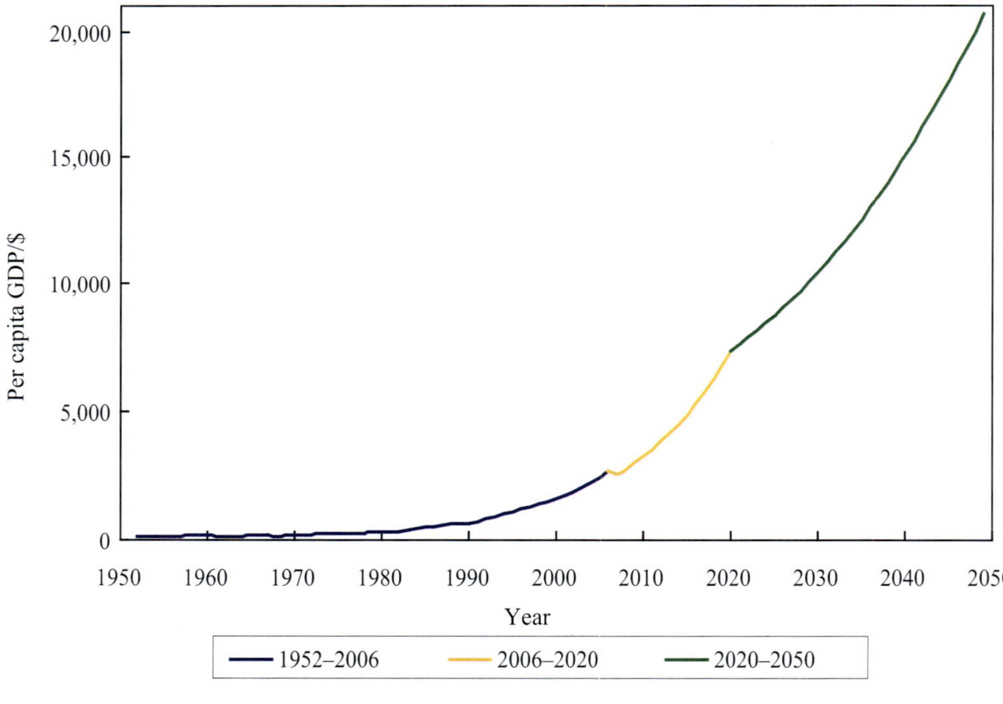

Fig. 3.3 Prediction of China's economic development to 2050

2. Upgrading of the Industrial Structure

China's industrial structure evolution from 1952 to 2006 is in line with the general law (Fig. 3.4): the proportion of primary industry in national economy is on decline, while those of the secondary and tertiary industries are increasing in correspondence. Since 1970, the secondary industry occupied a large proportion in national economy, larger than the tertiary industry on the whole; however, after 1980, the tertiary industry started to show a strong momentum with rising proportion in national economy. It is generally believed that China is currently in the middle stage of industrialization, which will continue to be the main driving force in China's economic development for a long period of time. In the next 20 to 30 years, China is expected to complete industrialization. Then the proportion of tertiary industry will keep on rising and become the mainstay of economic development.

The rapid economic growth and industrial change in China in the next 50 years suggests a more severe environmental challenge: on one hand, rapid economic growth will lead to more environmental pollution; on the other hand, to 2050, the rising output of the industrial sector, service sector and the transport sector also shows an industrial structure with more pollution. In technical aspect, although as intermediate input, the intensity of coal use is expected to drop, while the use of petrochemical products may increase and cause a substantial increase of air pollutants.

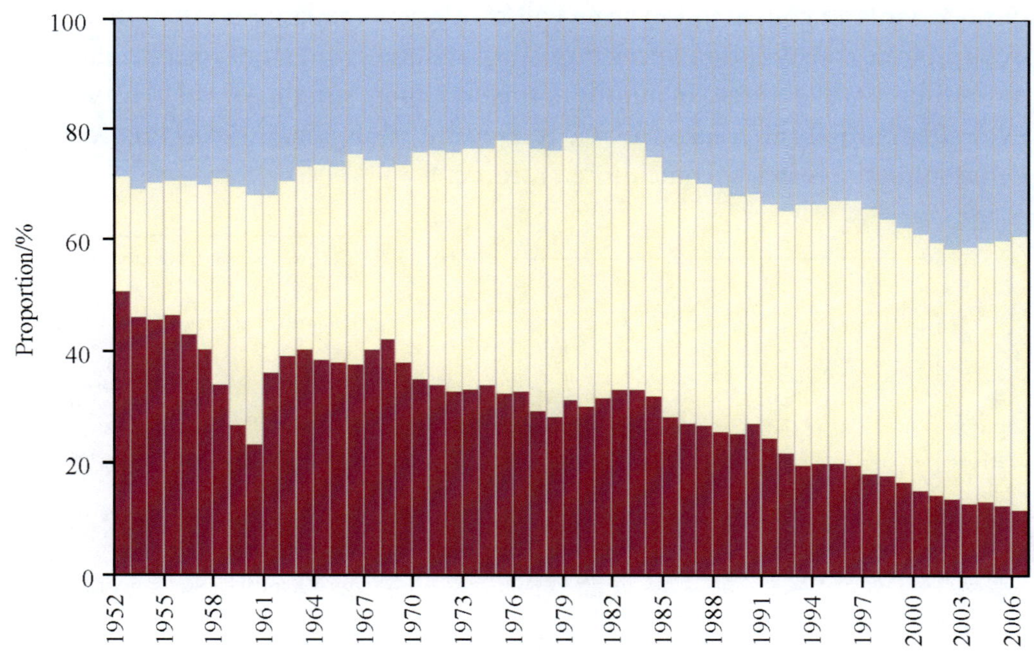

Fig. 3.4 The change of industrial structure from 1952 to 2005
Source: Fifty years of new China, the compilation of statistics (1949–1999), China statistical yearbook, 2007

3.4 Consumption Patterns

Fig. 3.5 shows the trend of changes since 1978 till now and in the future. It can be seen that no matter urban household or rural household, the income index is in an upward trend, almost without fluctuations. The net income index of rural resident was higher than the disposable income of urban resident before 2006. While Engel's coefficient manifests a certain degree of fluctuation in both urban and rural households, it shows a downward trend on the whole.

As for the future developing trend, both the net income index of rural residents and the disposable income index of urban residents will continue to rise, while the Engel's coefficient will keep on dropping in correspondence. In other words, the Engel's coefficient lowers along with the income growth. As a result, the residents will be prone to high-end consumption with the increased income and may show the trend of diversification. In terms of the consumption patterns in some developed regions, consumer durables, real estate, automobiles will become the major consumption objects for urban residents in future. The Engel's coefficient of rural residents drops slowly compared with that of urban residents, but the consumption pattern is gradually close to the urban residents. Such changes in consumption pattern may boost the industrial development thus further guide the industrial developing orientation.

In the next 50 years, on one hand, the majority of people will step into well-off living from poverty, plain food and clothing, hence the material production and consumption scale will continue to expand; on the other hand, the resource consumption keeps on rising, and the problem of insufficient back-up resource will emerge. In addition, the ever-increasing material consumption will substantially increase the waste amount, thus make the ecological and environmental issues become more prominent.

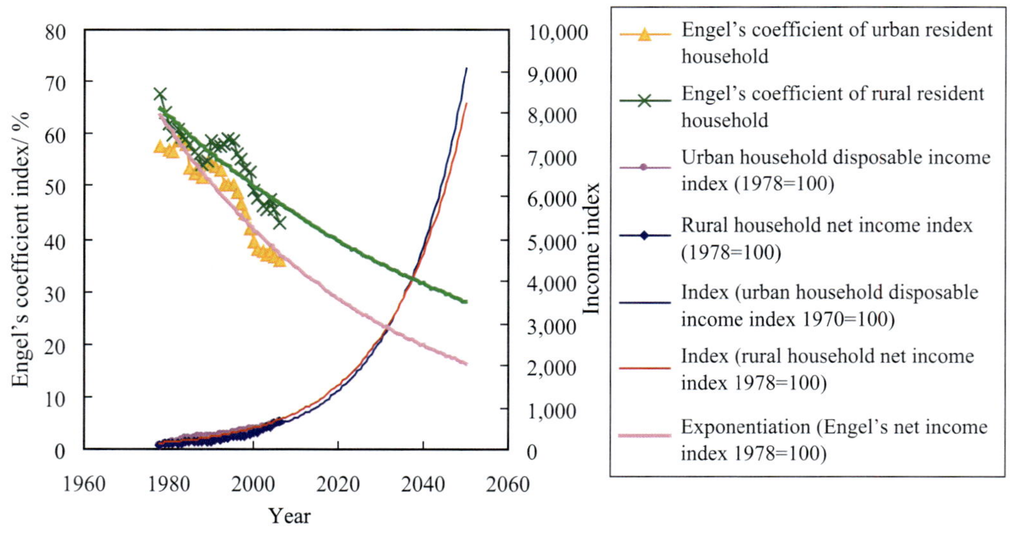

Fig. 3.5 The changing trend of consumption pattern in China

3.5 The Needs for Ecological and Environmental Science and Technology in the Next 50 Years

The rapid social-economic development in China has brought enormous social-economic benefits, but has also given rise to a great many environmental issues, some of which have reached staggering levels. Along with the population growth and social behavior evolution, the population will continue to increase in the coming decades and more resources will be consumed. Economic development and changes in consumption patterns (such as consumption of disposable product, buying cars and houses for comfort pursuit, etc) also bring increasing pressure to ecology and environment; urbanization process will be accompanied by material metabolism of per unit area, increasing energy consumption intensity, land-use change and the spreading infectious diseases brought by population concentration. Developed countries take advantage of their solid knowledge base, efficient innovation system to continuously produce knowledge and rapidly turn it into commercial value and social value, thus stay in the leading position in global knowledge society. During the process of catching up with developed countries, China attaches great importance to knowledge application and converts the external knowledge into its own wealth through strengthening knowledge digestion, absorption and re-innovation, so as to keep on shrinking the gap with developed countries.

Science and technology is the leading force in economic and social development. China's ecological and environmental technology development needs to grasp the general development trend in the world, to seize the major bottlenecks that constrain socio-economic development, and to further clarify the needs of scientific and technological development based on key technological fields that play a crucial and leading role in national long-term development.

1. Provide the Need of Scientific and Technological Support for Ecological Civilization and Construction of Ecological Engineering

Currently, China is in a crucial development period, facing a series of important tasks such as economic structure optimization, rational use of recourses, ecological environment protection, promotion of coordinated regional development, improvement of population quality and the total elimination of poverty, etc. The Third Plenary Session of the 16th Central Committee of the Party clearly put forward "Scientific Concept of Development" for the first time. The 17th National Congress of the CPC raised the strategic task of building "ecological civilization" and a number of major ecological projects needing to be activated. Realization of those strategies requires revealing the laws of spatial and temporal variation of China's ecology, environment and space, uncovering the evolution mechanism of man-land system and scientifically predicting the future development in a certain period, which calls for the support from dynamic monitoring data of all-directional, high spatial and temporal resolution ecological environment.

2. Provide the Need of Scientific and Technological Service for the Management and Optimized Allocation of Natural Resources as well as the Regional Sustainable Development

The issues such as supply and sustainable use of resources, environment protection and improvement, promotion of a virtuous circle of ecosystem, mitigation of the loss caused by disasters, coordination of the relationship between man and nature, are the fundamental and strategic ones faced by human society in seeking sustainable development in the 21st century, which are throughout the entire process of national economic construction and infiltrate in all aspects of human society development. The core to address those issues is the national program of optimized allocation and natural resource management to coordinate regional sustainable development, which requires the support of national-scale ecological monitoring and experimental data as well as the establishment of ecosystem experiment station in different regions to carry out comprehensive research and experiment of application basis and technology, to provide optimization models, key and supporting technology for local resource utilization, environmental protection and industrial development, and to conduct demonstration and promotion thus promote regional economic growth.

3. Provide the Need of Scientific Knowledge and Supporting Technology for the Construction of Ecological Environment

Ecological environment is the basic condition of human existence and development, also the foundation of economic and social development. According to China's ecological construction planning, by 2050, a virtuous ecosystem that basically adapt to the sustainable development will be set up; the soil erosion areas needing appropriate management will be mostly remedied; all the forest suitable land will be afforested with rational forestry species and tree structure; the forest coverage rate will reach and be stabilized above 26%; the "desertification, salinization and degradation" grassland will be completely restored. However, China has few original scientific and technological achievements in the academic field of ecological environment construction subject, with insufficient technological reserves, poor technology integration, high-level talents shortage; therefore, dynamic monitoring of ecosystem and environment evolution, dynamic analysis of driving force as well as revelation the mechanism of ecological process need to be carried out for a long period of time, so as to provide a strong scientific and technological support to address the major realistic technology issues in China's ecological environment construction.

4. Satisfy the Scientific and Technological Need of Coping with Global Environmental Change and Economic Integration

The developing trend of global environmental change and economic integration has made the national political and economic security become increasingly important. The present China needs to play a greater role in international affairs and assume the responsibilities in maintaining world political and economic order. We shall participate in the negotiation and decision of international rules and conventions, and also fulfill the obligations under international conventions. Global climate change and economic integration require the ecological environment study of China to fully utilize the favorable conditions provided by both domestic and foreign research developments as well as regional advantages of our country for the integrated monitoring and scientific research of ecological process and change patterns in different regions and on different scales, so as to make a unique contribution to the research on global environmental change and sustainable development.

5. Provide Scientific and Technological Support for the Fulfillment of International Conventions and Cooperation

Various environmental issues caused by global environmental changes have become a key topic in international diplomacy with an increasingly prominent status. At the same time, the global environmental issues and the countermeasures play an increasingly important role in national economic development. China is the largest developing country in the world with the environmental issues receiving wide international concerns.

Presently, China has signed over 30 international environmental treaties, conventions and agreements (not including bilateral treaties and agreements, etc), with the content covering global climate change, ozone layer protection, biological diversity, wetland conservation, desertification control, marine environmental protection, international species trade and hazardous waste control, etc. In those diplomatic activities, Chinese government needs to take the comprehensive, detailed scientific observation and research data both at home and abroad as the basis.

6. Accumulate Scientific Data for National Long-term Spatialization, Promote the Need of Scientific Development of Earth System

China's complex and diverse natural environment has formed a number of distinctive geographical-ecological units (such as the desert, the cryosphere, the Loess Plateau, karst areas, inland rivers, farming-pastoral zone and the arid valley, etc). They are geographical areas bound by the natural conditions and facing with difficult problems, also the hot areas researched by earth system science and the major objects of earth system science with studying the interaction among environment, ecology, disaster and human spatial behavior as the major task. To systematically carry out observation and research of geographical characteristics-ecological unit in different regions of China, and to integrate all types of ecological environmental information for the accumulation of long-term spatialization scientific data resource, not only address the need of solving the major resource and environmental issues constraining China's social-economic development, but also the need of accumulating national data resource, promoting the development of earth system science and technological innovation.

7. The Technological and Strategic Need for Environmental Security

China's overall ecological environment has entered a new stage of large-scale ecological degradation and composite environmental pollution. The environmental issues occurred in a century-long industrialization process in phases of the developed countries are collectively emerging in China. The environment is facing growing pollution pressure and the environmental pollution trend will also largely depend on the mode of economic and social development, environmental pollution control strategies, environmental management levels, technology levels of pollution control and the implementation degree of control measures, etc. Meanwhile, the issue of environmental pollution is lacking of systematic, coordinative and innovative scientific research, the technical support for environmental pollution control is relatively weak and the regulatory platform still needs improvement. The potential hazard of environmental status dropping with rapid socio-economic development or even occurring mutation is becoming severe. It can be expected that in next 5 to 15 years, or even longer, along with China's rapid socio-economic development, the restraining and threatening "bottleneck" impact of environmental pollution will become increasing serious. In face of the severe status quo and developing trend of China's environmental pollution, pollution control, treatment, rehabilitation and environmental security protection will become one of the important national technological and strategic needs of environmental pollution control.

8. The Technological and Strategic Need of Maintaining Ecosystem Health

China's environmental and socio-economic background has changed dramatically and will accelerate along with the economic process and urbanization process. Natural ecosystems are undergoing profound and far-reaching changes. At present and in a long period in future, the major threats faced by China's ecological environment include: ① high environmental risk brought by chemical substances that continuously and stably existing in ecological environment; ② biodiversity loss and declining ecosystem service function. In face of China's severe status quo and development trend of environmental pollution, especially the complex migration process, pollution effect and representation behavior after the immigration of great deal of chemicals, there is still no breakthrough in the areas of ecological health and environmental security, which can not provide effective technical support for the restoration of degraded ecosystem of multi-scale pollution characteristics, transmission laws and control indexes. In a long period of time in future, the scientific and technological strategy of ecosystem health centered by environmental security and risk control will be a key national strategic need in environment pollution control.

9. The National Need for Prevention and Mitigation of Weather and Climate Disasters

China is located in East Asian monsoon area, adjacent to the Pacific in the east and Qinghai-Tibet Plateau in the west, with a quite complicated climate spatial distribution, including the typical arid area mainly consist of Gobi and desert, the semi-arid zone dominated by loess plateau and grassland in middle east area as well as the humid zone in the east. Impacted by the factors such as geographical location and the unique topography and geomorphologic conditions, China is a sensitive and vulnerable region in global climate changes.

Since the late 1970s, under the circumstance of exacerbated global warming, China has endured a huge climate change in the form of frequent weather and climate disasters such as large-scale drought, floods, heat, continuous heavy snow and freezing weather, etc. The concrete examples include the persistent drought in North China since 1976 and the huge floods in Yangtze River valley in the summer of 1998, etc. In recent years, the area of arable land suffering from drought and flood was over 500 million mu annually, resulting in about 20 billion tons' grain output reduction; the economic loss in 1990s was as high as 200 billion yuan, accounting for about 3% to 6% of the national GDP at that time. Since entering the 21st century, the weather and climate disasters in China shows an obvious aggravating tendency, bringing more serious loss to the national economy, such as the grave drought in Chongqing in 2006, the serious flood in Huaihe River basin and torrential rains in Sichuan, Chongqing and Guangdong, etc, in 2007 as well as the severe freezing rain and snow in south area in the beginning of 2008, etc. Those weather and climate disasters closely related to the global climate change has caused huge losses to industrial and agricultural production, national economy and people's lives and property, and brought about a series of severe environmental and social issues.

4 The Trend of Global Ecological and Environmental Changes

4.1 Climate Change

The Intergovernmental Panel on Climate Change (IPCC) established by World Meteorological Organization and the United Nations Environment Program launched the Fourth Assessment Report in 2007, clearly pointing out that human activities had a huge impact on global climate, particularly since the middle of last century. In addition, on the basis of the risen global surface temperature, that is, 0.74 ± 0.18°C from 1906 to 2005, the report forecasted at the end of the 21st century, the global climate would have another rise of 1.1°C to 6.4°C. Current policies and the trend of greenhouse gas emission will lead to a rapid rise in global temperature. It was estimated that in 2050, the greenhouse gas emission would be 37% higher than that in 2005 and would produce a series of far-reaching impacts. Therefore, the emission trend should be reversed to protect the climate and make the greenhouse gas emission in 2050 significantly lower than the current level. The primary driving force to emission growth consists in using fossil fuels as well as implementing non-sustainable land utilization policies, such as deforestation. By 2050, agricultural emission and waste emission will greatly contribute to the growth of greenhouse gas emission.

The Observed Climate Change

Climate change is beyond doubt, which can be detected from the current observation of rising global average temperature, sea temperature, sea level, as well as large-scale melting snow and ice.

In recent 12 years (1995–2006), 11 years are among the top 12 warmest years according to the global surface temperature measuring data (since 1850). The linear trend of temperature of the past century (1906–2005) is 0.74°C [0.56°C to 0.92°C] (the figures in the square brackets show the possibility of a certain optimal valuation is a 90% uncertainty interval, that is, the possibility for the value to be greater than the scope given in the bracket is 5%, while the possibility of being less than the scope is 5%; uncertainty interval is not necessarily corresponding to the symmetrical value of a certain optimal valuation). This trend is greater than 0.6ºC [0.4ºC to 0.8ºC], the trend (1901–2000) given by the Third Assessment Report (TAR). There is a general rise in global temperature, with a larger rising margin in high latitude region of Northern Hemisphere and a faster land warming rate.

The gradual increase in sea level is consistent with the warming. Since 1961, the average sea-level rising rate was 1.8mm [1.3mm to 2.3mm] annually, while the average rate since 1993 is 3.1mm [2.4mm to 3.8mm]; thermal expansion, the melting of glaciers, ice caps and polar ice caps have contributed to the rising of sea level. There is no clear conclusion on whether the accelerating rising rate of sea level from 1993 to 2003 reflects decadal variation or longer-term upward trend.

The observed reducing snow and sea ice area is also consistent with the warming. Satellite information since 1978 shows the average Arctic sea ice area has been shrinking at the rate of 2.7% [2.1% to 3.3%] every decade and even higher in summer, which is 7.4% [5.0% to 9.8%]. The average area of mountain glaciers and snow cover in both hemispheres has manifested the shrinking tendency.

From 1900 to 2005, there was a significant precipitation increase in the east part of North and South America, Northern Europe as well as the north and central part of Asian region; while it decreased in the Sahel, the Mediterranean, Southern Africa and South Asia. Globally speaking, since 1970s, the area affected by drought *may* (italics refers to the reviewed statements of the relevant uncertainties and credibility) have expanded.

It is most probably that during the past 50 years, the frequencies of cold day, cold night and frost in most land areas have declined, while hot day and hot night occur usually. The occurrence of heat waves has been frequent in most land areas and strong precipitation frequency has increased in a majority of areas. Since 1975, in global context, the extremely high sea level events (except for tsunami, which is not among the events caused by climate change; extremely high sea level depends on the mean sea level and regional weather systems; in this report, the definition is, at a particular base period, 1% of the highest per hour sea-level value observed by a certain station) have increased.

Observation evidence (largely based on various data sets of the entire period since 1970) shows that since around 1970, the strong tropical cyclone activities have increased in North Atlantic, but there is limited evidence of such phenomenon in other parts of the world. No clear trend can be found in the annual occurrence of tropical cyclone. Currently, it is still difficult to determine whether the cyclone activities have a longer-term trend of change, especially before 1970.

In the second half of the 20th century, the average temperature in Northern Hemisphere was very likely higher than any other 50-year-period in the past 500 years, and likely the highest in at least the past 1,300 years.

4.2 Air Pollution [10]

It is estimated that by 2030, urban air quality will be even worse. In most regions, the concentration of respirable particles (PM^{10}) in the air has started to exceed the target level. In OECD countries, the major source of man-made air pollution is still road transportation and the use of fossil fuels. In many developing countries, wood burning also contributes to air pollution. In recent decades, a majority of developed countries have reduced air pollution and achieved decoupling from economic growth. However, pollution from other countries is gradually undermining the local urban air quality management thus turns it into an international problem. Moreover, the management of air pollution caused by aviation and narigation is lagging behind that of road traffic. Therefore, in response to air pollution, more attention should be paid to marine transport and the disposal of air pollutant precursor on the ground. And it is quite important to consider the transmission of air pollution from one continent to another in formulating national air quality policies.

The Imminent Control of Air Pollution

Air pollution is one of the most serious global environmental problems. Currently, the developing countries are facing with almost the same air pollution issues encountered by developed countries since the middle of 20th century. *Air Pollution's Causes, Consequences and Solutions* issued by World Resources Institute analyzed the status and challenges of air pollution faced by the the world, especially in developing countries such as China, from the perspective of developed countries.

Outdoor air pollution, regardless of the visible haze or invisible ozone and carbon monoxide, is a problem existing in almost all countries.

It is estimated that the annual death caused by only outdoor air pollution is 800,000. Moreover, indoor pollution has resulted in the premature death of 1.6 million people. In many urban areas, especially in developing countries, air pollution is the primary environmental threat to human health.

However, some cities in developed or developing countries once being heavily polluted are able to significantly improve air quality and control it within the safety level. The successful experience of those cities usually derives from perfect combination of strict standards, reasonable laws and regulations and comprehensive traffic solution program.

The Types of Air Pollution

It is tested that three kinds of widespread air pollutants are most hazardous to environment and human health: ① sulfur dioxide, generates from the combustion of high-sulfur fossil fuels in smelting or other industrial processes; ② nitrogen oxides, generates from mobile sources (such as motor vehicles) and stationary sources (such as power plants); ③ particulate matters, refer to the small particles suspending in the air, which is classified in accordance with the particle size; the particle with the size less than 10μm is called PM_{10} (respirable particulate matter), and called $PM_{2.5}$ (fine particulate matter) with the size below 2.5μm; automobiles, power plants and industrial processes can generate particulate matters.

Other pollutants include: ① carbon monoxide, an odorless gas, mostly emitted by the motor vehicles without catalytic converter; the combustion of fossil fuel in industrial production will also emit some carbon monoxide; ② lead, a kind of particulate matter emitted by the motor vehicles using lead gasoline; ③ ozone, generated by volatile organic compounds and nitrogen oxides in certain weather conditions (sunny and windless days); ④ volatile organic compounds, including hydrocarbons, ethanol, acetaldehyde and ethyl ether; among which, hydrocarbons, playing a role in ozone formation, which generate from industrial process and automobile exhaust.

Air Pollution in Global Context

In the United States and other developed countries, nitrogen oxides and sulfur dioxide emissions are usually related to power plants using fossil fuels, dropping along with the increasing power generating capacity. The level of nitrogen oxides and sulfur dioxide in car exhaust has also dropped to a certain extent, although the total number of vehicle miles in the United States almost increased 150% since 1970. These results are largely attributable to more strict motor vehicle engine and fuel standards.

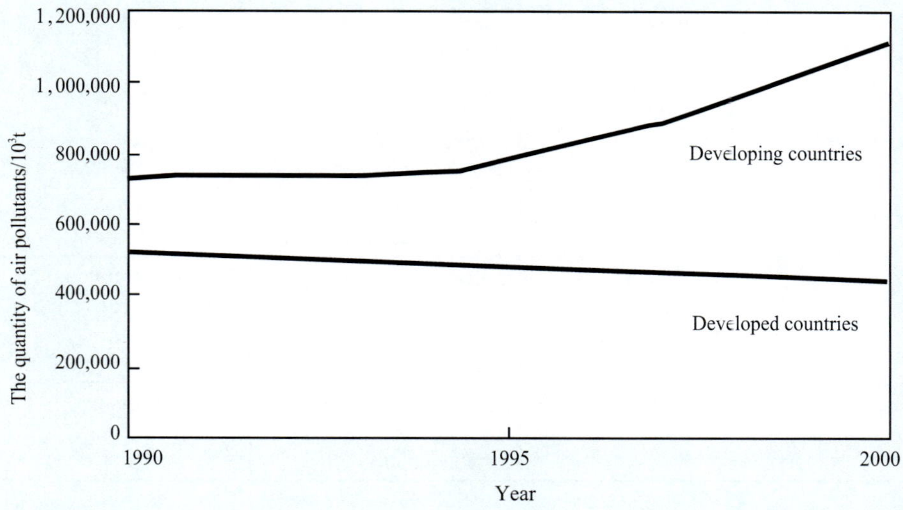

The situation of air pollution in developed and developing countries from 1990 to 2000

However, in most developing countries, the growing number of automobiles, power plants and factories has not brought cleaner technology or more strict regulations. From 1990 to 2000, air pollution in developing countries increased by 50%. Air pollution has become a prominent issue in Asian, and the death caused by air pollution accounts for 65% of the pollution-related death.

The Impact of Air pollution on Health and Environment

The adverse impact on health of long-term living in an environment of air pollution has been well confirmed. A research of World Bank found that when every 30μg/m^3 was added to daily PM_{10} concentration, the signs of respiratory disease would increase 9%. Such growing trend has been found in both high-exposure group (those living or working near the factory or road) and low-exposure group living in a better protected environment.

PM_{10} is also related to 1% to 2% of cardiovascular disease mortality increase and 3% to 6% of respiratory disease mortality increase even if adding only 10μg/m^3. Other studies also associate air pollution with the low birth weight of newborn babies.

Air pollution will severely damage the natural environment. Sulfur dioxide can lead to acid rain, thus harm the lakes, rivers and forests. The active nitrogen in the emission deposit of nitrogen oxides can cause excessive nutrients in terrestrial ecosystem, and the generated runoff can result in eutrophication in the coastal zone. In the eastern part of the United States from 1970s to 1980s, strict power plant regulations were enacted due to the growing concern on air pollution and acid rain.

The Main Source of Air Pollutants and the Impact on Health and Environment

Pollutant	Main source	Impact on health	Impact on environment
Sulfur dioxide	Industry	Respiratory disease, cardiovascular disease	Acid rain precursor, a threat to lakes, rivers, trees and cultural relics
Nitrogen oxides	Automobile, industry	Respiratory disease, cardiovascular disease	Nitrogen deposition leading to excessive nutrients and eutrophication
Particulate matter	Automobile, industry	Particles enter deep into the lung and can enter into the blood	Poor Visibility
Carbon monoxide	Automobile	Headache and fatigue, especially to those in Poor cardiovascular condition	
Lead	Automobile (using lead gasoline)	Accumulated in the blood, damage the nerve system	Fish / animal death
Ozone	Generated by nitrogen oxides and volatile organic compounds reaction	Respiratory disease	Reducing grain output and forest growth; smog precursor
Volatile organic compounds	Automobile, industry process	Eye and skin irritation, nausea, headache and carcinogenicity	Smog precursor

Measures for Improving Air Quality

In the United State, Clean Air Act of 1970, a milestone document, established National Ambient Air Quality Standards, which is based on the impact of main air pollutants (sulfur dioxide, nitrogen oxides, PM_{10} and $PM_{2.5}$), as well as lead, ozone, carbon monoxide on human health and environment. Many parts of the United States often meet or exceed the standards of those pollutants. It is worth noticing that the exception is the ozone concentration–on average, it is reducing slowly, but in some areas, it has increased.

Other countries (such as China) are still making their efforts to achieve their own national standards that are usually less strict. A government report in 2007 concluded that 48% of China's urban residents lived in the cities failing to meet national air quality standards. It is estimated that only coal burning has resulted in

50,000 deaths and 400,000 cases of chronic bronchitis. World Bank predicts that by 2020, China will spend 390 billion yuan (equivalent to 13% of the estimated GDP) on medical service of diseases caused by coal combustion.

While some of the cities in developing countries are still trying to control the deteriorating environment, some are able to control (or at least reduce) the air pollution. For example, the level of pollutants in Bangkok can reach the standards of the United States, and is close to European standards, while Bangkok was infamous for severe air pollution 15 years ago. The success of Bangkok in the control of air pollution is partly due to the luck and its unique geographical characteristics and industrial structure, but most of the reasons are attributed to innovative solutions.

The methods of Bangkok, namely, the sensible regulations like emission standard of exhaust pipe, and unitary repairing measures such as street sweeping and massive expansion of rail traffic, have become a model of air quality management. Although the circumstances of different cities are different, the experience of Bangkok air pollution control is applicable anywhere. Integration is the key; the issue can not be addressed by one method independently. Urban air pollution is a complex, multi faceted issue, posing challenges to health and the environment, transportation and engineering, economic and policy.

The problem in Beijing clarified the need of integrated solution for controlling air pollution. Even though the Chinese government has made greatest effort (namely, the implementation of "odd-even" traffic restriction), the air pollution in Beijing still can not be effectively controlled. Although automobile is the main source of local air pollution, the polluting factors out of Beijing are actually the most important determinants in urban air pollution, namely, volatile organic compounds emitted from the small factories in the neighboring Hebei Province, as well as the particulate matters produced by large-scale coal mines and coal-fired power plants in Shanxi Province.

Therefore, the solution to air pollution in Beijing may be more complicated than simple traffic restrictions on automobile or the emission of exhaust pipes. The actual potential driving factor of the problem is another issue, which needs the cooperation among cities, industries, and between the local government and the central government.

A Practical and Innovative Solution

No matter the solution is voluntary, mandatory or incentive-based, it is required to be practical and innovative. Those solutions also need adjustment to conform to the special geographical and economic environment of the city. When those measures are under the supervision of knowledgeable environmental managers, who are also the executors committed to good environmental management, those measures may still have a long way to go, so as to reduce the harmful impact of air pollution on health and environment.

Source: Zeng Jingjing. 2008. The Imminent Control of Air Pollution. Science News, 19: 26–27

4.3 Biodiversity

It is predicted that by 2030, biodiversity loss will continue, especially in Asia and Africa. The reasons for biodiversity loss mainly include land-use change, unsustainable use of natural resources, invasion of alien species, global climate change and pollution. In past few decades, the number of natural reserves had a significant increase, but the marine natural reserves were less representative in all kinds of reserve types. In the future, as a result of the accelerating expansion of agricultural and urban land-use, the natural reserves for biodiversity conservation will play an increasingly important role.

4.4 Freshwater

It is estimated that by 2030, a population of over 3.9 billion (approximately accounting for 47% of the world's total population) will live in areas of serious water shortage, especially in non-OECD countries; more than 5 billion people (about 67% of the world's total population) will be lack of public drainage facilities, 1.1 billion more than present; in coastal waters, there will be near 55 million tons' nitrogen come from inland, 4% higher than that in 2000; the land area with higher risk of erosion will increase 1/3 or more, reaching 27 million km^2 (about 21% of the world's total land area). Effective policy will be helpful to the solution of main water resource challenges, Therefore, progress will need to be made in the following aspects in future: integrate water resource management into the departmental (such as agricultural) and land-use policies; ensure stable application of Polluter Pays Principle (PPP) and the User Pays Principle (UPP) through price regulation and control; reduce the water problems by means of reducing subsidies.

4.5 Waste and Logistics

Illegal shipment and improper management of waste materials and products have posed a huge risk to human health and environment. In next few decades, the management of rapid increasing urban waste will become a great challenge to non-OECD countries. Despite the urban waste in developed countries is still increasing, the growth rate slows down, which has been basically decoupled from economic growth. With the continuing growth of the global demand for raw material and the waste production and handling capacity, the traditional unitary policy of waste may be insufficient to improve the utilization of raw material and make up for the impact on environment generated during the production and utilization. The new integrated approach can balance the impact on environment throughout the life cycle of raw material waste. In this approach, more attention is paid to the utilization of raw materials, product design and reuse, waste prevention, recycling of end-of-life raw materials and products, as well as the environmentally sound management of residual substances.

4.6 Environment and Health

If there are no more strict policies to better address the issue of environmental pollution, the negative impact of air, water and food source pollutants on human health will continue increasing. For example, it is predicted that from 2000 to 2030, the number of premature death caused by ground ozone and PM_{10} will increase. Tens of thousands of chemicals invented by human beings will also bring tremendous threat to ecological health. The toxic pollutants in chemicals are not only poisonous, but contain many other effects at the same time, such as carcinogenicity, teratogenicity, mutagenicity and also with endocrine disrupting effect, directly threats the survival and reproduction of wild animals and even human beings. In the next few decades, the monitoring, evaluation and management of toxic pollutants will become a key link of ensuring the health of human body and ecological environment.

The research on environment and health mainly involves the environmental chemical behavior and exposure pathways of pollutants as well as the toxic effects. The toxicity evaluation of chemicals is the basis of the study and an important link of protecting human and ecological health. For instance, on Dec 18, 2006, European Parliament and Council of the European Union formally passed the act of Registration, Evaluation, Authorization and Restriction of Chemicals (REACH), aiming at establishing an unified evaluation system of the hazards of the new and existing chemicals. REACH puts forward the requirements of the toxicity and ecological security to the entering tens of thousands of chemicals. In face of the REACH act challenge, China is lacking of the methods of chemical eco-toxicological evaluation and the relevant standards are almost in blank. Hence, China is in urgent need to carry out the development and standardization research of the methods of chemical eco-toxicological evaluation, so as to positively respond to REACH act; meanwhile, strengthen the evaluation of the eco-toxicological impact of imported chemicals on China's biology to protect the health of people and ecological security in China.

Registration, Evaluation, Authorization and Restriction of Chemicals

In order to address the practical issues existing in chemical safety management and overcome the drawbacks of the current management mode, through extensive research and seeking the views of stakeholders by the competent authorities of EU chemicals, European Commission issued the act of Registration, Evaluation, Authorization and Restriction of Chemicals (REACH). EU looks forward to strengthening the protection of human health and environment, changing management focus and ways through the new act to raise the supervision intensity on the chemicals impacting health and bringing environment risks. It is expected that the act will exert an influence on all chemical involved productions and (or) import in EU after implementation.

REACH is established on the basis of self-regulation and it holds that chemicals can only be put into production (use) after the information related to their nature (physical, toxicity, environmental impacts, etc.) being fully comprehended.

Firstly, implement unified data requirements and application registration procedures on the new and existing chemicals, which overcomes the management deficiencies of current legislation that focusing on new substances but in negligence of the existing ones, and gradually solves the deficiencies in existing substance security management as well as fills the blank of security data insufficiency.

Secondly, the emphasis of management carried out by the competent authority will be transferred to the three chemicals that having the greatest impact on health and environment [CMR (carcinogenicity, mutagenicity, reproductive toxicity) substances, PBT (persistent, bio-accumulative and toxic) substances and vPvB (very persistent and very bio-accumulative) substances, etc]. The production, import and use of those substances must be authorized and approved by European Commission. The protection of human health and environment can be maximized through strengthening the management of those substances. Those measures will bring long-term benefits to the workers in the workplace, to public health as well as to the environment, thus greatly reduce the cost caused by future disease treatment and disposal.

Thirdly, the act clarifies the responsibilities of the chemical industry (corporation) being responsible for providing chemical security data and assuming preliminary evaluation. The enterprises' responsibilities of identifying chemical risk characteristics and initial risk evaluation, or affording some reasonable monitoring cost of the prevention and control of chemical pollution reduce the cost burden of the competent authorities in risk investigation and management.

Fourthly, the act extends the chemical safety responsibilities to the downstream users of the supply chain, which improves the spreading of chemical security information on the entire supply chain.

Fifthly, because the act stipulates that the registered chemical security data can be shared and the cost can be jointly afforded, the times of animal experiment will be reduced, which also achieves the purpose of animal protection.

Sixthly, the implementation of the act restricts the production and use of the three most dangerous chemicals, which will promote and encourage the chemical corporations to eliminate old products and develop new alternatives as well as conduct technological innovation.

REACH act reflects the focus of international chemical security management. It clearly points out that the production, sales and use of the following three highly concerned substances can only be conducted after gaining the approval: substances conforming to the first or second type of classification based on the carcinogenicity, mutagenicity and reproductive toxicity (CMR substances); substances with persistent, bio-accumulative and toxic characteristics (PBT substances) and substances that are very persistent and very bio-accumulative (vPvB material); substances have quite serious and irreversible effects on human health and environment.

REACH act changes the focus of chemical security and environment management and invests the limited resources of competent authorities into the examination and approval of the confirmed three substances, namely, CMR substance, PBT substance and vPvB substance, which has full scientific nature and rationality.

REACH act shows the principle of chemical management based on risk management. The act stipulates the identification standards, examination procedures of the above-mentioned three substances. It also proposes a principle, based on whether the risk of one chemical to health and environment is acceptable or controllable. Even if the risk can not be properly controlled, as long as the registration applicant can demonstrate that the socio-economic benefits it brought will be far greater than its risk to human health and environment, and there are no appropriate substitutes or alternative technologies currently, then the registration of such substance also has the possibility to win the approval. The principle fully reflects the risk-based social acceptability and social cost-effectiveness; the chemical security management is not conducted just in accordance with the inherent risk of chemical properties or the degree of the danger. The management idea is reasonable and also in line with the development trend of international chemical security management in recent years.

Source: Environment Department, Enterprise and Industry Office, European Commission. 2006. REACH SUMMARY

5 China's Ecological and Environmental Characteristics and the Evolution Trend

As a developing country enjoying rapid development, China has its own characteristics in ecosystem service and environmental quality in addition to the unique features in socio-economic development. Moreover, there is a big gap between China and the developed countries in America and Europe in terms of ecological and environmental technologies. Hence, in the formulation of Roadmap for Development of Ecological and Environmental Science and Technology in China, it is necessary to draw on the science and technology development strategies of the developed countries on the one hand, and have a bold foresight to avoid endless copying on the other hand. Therefore, we should also take the leaping development of science and technology into consideration and strive to become a leading force of world ecological and environmental technology development in a number of fields.

Of course, what is the most important in roadmap formulation is to focus on the characteristics of ecology and environment, analyze China's long-term development program and the trend of socio-economic, technological progress and global environmental changes as well as predict China's ecological and environmental evolution trend in the next 50 years, so as to put forward a strategic, directional and operational roadmap of science and technology development.

5.1 Biodiversity Is Being Destroyed and Wild Animals and Plants Are Decreasing

As a result of natural factors, such as climate change, and historical man-made destruction like deforestation, grass land reclamation, and indiscriminate hunting and fishing, the current trend of China's biodiversity loss has not been effectively controlled. Many ecosystem functions continue to degrade, the proportion of endangered species keeps on increasing, the invasion of alien species is aggravating and the loss of genetic resources and erosion are exacerbating.

Presently, China is in a period of rapid population growth and economic development. The biodiversity has been severely threatened at three levels, namely, ecosystem diversity, species diversity and genetic diversity. The endangered species criteria represents the priority of protection sequence. In November 1994, the 40th Council Meeting of the International Union for Conservation of Nature (IUCN) formally employed the revised Mace-Lande Endangered Species Criteria as its own criteria. Although the *IUCN Red List* of endangered species not belongs to international laws or state laws, it has a far-reaching influence on the decision-making of intergovernmental organizations and non-governmental organizations as well as the formulation of natural laws and regulations of each country; it also plays a role in the theoretical research of conservation biology. In researching the formulation of China's endangered wildlife criteria, we found there would be some problems if directly employ IUCN endangered species criteria to assess the level of endangered species, such as: ① How to distinctively treat the species in small number and with narrow distribution range from those with sharp drop in number and shrinking living space due to human activities? ② Can the same endangered criteria be employed in different animals? ③ How to distinctively treat the different species quantities and densities in edge distribution area and core distribution area? ④ How to deal with the local species extinction and local endangerment? ⑤ Some species are endangered in the wild, but can be saved through artificial propagation, how to deal with those endangered species? ⑥ If there is no accurate historical information and statistical data of the species and their habitats, how to employ the endangered species criteria to assess the degree of their endangerment? Based on the principles of simple and practical, not only being in line with the existing international criteria but also conforming to domestic levels of wild animal protection, being able to make assessment and grading by utilizing the existing literature and data both at home and aboard, as well as being convenient for species resource classification management, we should combine the national conditions on the basis of IUCN Red List Criteria to develop a set of endangered animal grading methods for distinguishing the living histories of wild animals [11].

The *China Species Red List* published in 2004 employed *IUCN Red List Categories and Criteria* of the version in 2001 to assess the extinction risk on 10,211 kinds of plants and animals in China. China's situation of species endangerment is far more serious than the pass, with the proportion of endangered species generally from 20% to 40%; especially, as for the higher plants, the endangered species or those near endangerment are approximately accounting for 15% to 20% of the total species, higher than the average level of the world. The biodiversity drops sharply and around 1/5 of higher plants are severely threatened; among the 1,121 world endangered species listed in Convention on International Trade in Endangered Species of Wild Fauna and Flora, 190 species are in China. The invasion of alien species poses a threat to ecological safety, causing huge economic losses.

Biodiversity

With rapid population growth and exacerbating human economic activities, biodiversity, the most important basis for human survival, has been seriously threatened. China is one of the countries with extraordinarily rich biodiversity and also among the list of countries with severely threatened biodiversity.

Biodiversity is the feature of diversified entity group. Entities at every level–genes, cells, populations, species, communities and even ecosystems are more than one category, that is, diversity is everywhere. Therefore, diversity is the basic characteristic of all living systems. Biodiversity includes all plants, animals, microbial species as well as all ecosystems and their ecological shaping processes. It is a concept with wide content that describing the extent of natural diversity and also a function of time and space. Biodiversity is shown in each organizational level of living system, from gene to ecosystem, mainly including three levels, namely, species diversity, genetic diversity and ecosystem diversity.

Genetic diversity refers to species genetic changes, including genetic variation in quite different populations within a species and the genetic variation within the same population. Species diversity is the most important source of various diversities above the species level. Genetic variation, life history characteristics, population dynamics and genetic structure, etc., determine or influence the ways of interaction between one species with other species. Moreover, the diversity of species is the determinant of a species to successfully respond to human disturbance. The degree of intra-specific genetic variation also determines the evolutionary potential. All genetic diversities take place at the molecular level and are closely related to the physical and chemical properties of nucleic acid.

Species diversity refers to the biological diversity of species level. It is a regional species diversification, mainly studying the species status in a certain region from the perspective of taxonomy, systematics and biogeography. The status quo of species diversity (including the status quo of being threatened), the formation, evolution and the maintenance mechanism of species diversity, etc, constitute the major research contents.

Ecosystem diversity refers to the diversification of biosphere habitats, biological communities and ecological processes, as well as the amazing diversity of habitat differences and ecological process changes. Habitat diversity is the basic condition for forming the diversity of biological communities and even the entire biodiversity.

The diversity at each organizational level is quite important. Each biology branch relevant to the diversity at different levels has its own research focuses and unique research approaches. Species diversity is the basis of biodiversity study and ecosystem diversity is the focus of biodiversity study.

Source: Ma Keping. 1993. On the Concept of Biodiversity. Biodiversity, 1 (1): 20–22

China Species Red List

China Species Red List is composed of 6 volumes. The first volume is the *Red List* of about 10,000 species of plants and animals, including the analysis of endangered species in China, the translation version of *IUCN Red List Categories and Criteria* and the *Application of IUCN Red List Criteria at Regional Levels*, the names of about 10,000 species (scientific name, Chinese name English name and Chinese local name), the proportion of the distribution in China, the assessed level of endangerment as well as the criteria and reason, and IUCN global assessment grade, etc. The latter 5 volumes are Vertebrate Volume, Invertebrate Volume and Plant Volume (3 volumes), the content of which is the Species Databank for Conservation of the approximate 6,000 threatened species distributed in China. The up-to-date information of each species has been collected, including classified information (species scientific name, Chinese name, English name, important synonym and classified information remark), the existing endangered grading information (*IUCN Red List*, *China Red Data Book*, CITES appendix and national protection level), the global proportion of the distribution in China, the criteria and reasons based on the endangerment category assessment, habitat, distribution (distribution map), population status, risk causing factors, protective measures, reference literatures, assessment personnel and time, etc, which is of important academic reference value.

The publishing of *China Plant Red Data Book* (volume 1) and *China Red Data Book of Endangered Animals* (4 volumes) since 1980s to 1990s, was warmly welcomed by people from the circles of scientific research, management, protection and law enforcement and received a strong social response.

In late 1990s, the International Union for Conservation of Nature (IUCN) issued new criteria of species endangerment category, applicable to all the higher and lower plant and animal species and launched the CD of *Global Species Red List*. Subsequently, all countries carried out assessment of their own species status in accordance with the new criteria. Because the *Global Species Red List* is not usually identical with the specific country or regions, the information collection can not be comprehensive.

Therefore, since the late 1990s, the Biodiversity Working Group of China Council for International Cooperation on Environment and Development initiated and organized the *China Species Red List* project aiming at assessing the status quo of species conservation. It specially organized some core experts engaging in different categories nationwide to assume the species assessment of the relevant categories. The hosts of *IUCN Red List* project were invited to join the training and seminars in China at the project starting phase. The project lasted for 3 years, which made a comprehensive assessment of the species of animals and plants situation in our country by applying the new IUCN criteria. Quite a few seminars were held to conduct strict and scrupulous review and supplement for the first draft, so as to ensure the academic level; at the same time, the latest international classification system was adopted for a number of categories or systems such as birds and mammals to keep in line with international standards.

The book has demonstrated a historical species assessment in China, reflected the status quo of species survival in the turn of the century, shown the achievements and deficiencies of the protection for over 50 years and pointed out the critically endangered species and endangered species needing protection in priority. The main content of the book is in both English and Chinese and uses internationally accepted markers so as to be convenient for both domestic and foreign readers.

Source: Song Wang, Yan Xie. 2004. China Species Red List. Beijing: Higher Education Press

5.2 Accelerating Land Desertification and Serious Ecological Hazards Pose a Threat to National Arable Land

According to the estimation of United Nations Environment Program, 1/4 of the global land is threatened by desertification. More than 250 million people in the world are directly suffering from desertification; meanwhile, because of the leanness of arable land and pastures, the livelihood of more than 1 billion people in over 100 countries is in dangerous situation. Drought could trigger desertification, but the activities of human such as over-cultivation, overgrazing, deforestation, poor irrigation, etc, are usually the main incentives. *United Nations Convention to Combat Desertification in Those Countries Experiencing Serious Drought and/or Desertification, Particularly in Africa* (referred to as "the convention") aims at dealing with this issue. The convention signed by 172 countries provides a basic framework to all activities of combating desertification. The convention focuses on the improvement of land productivity, land transformation, as well as the conservation and management of land and water resources. It emphasizes the mass participation, lays emphasis on creating a "favorable environment" for the local people so as to enable them to reverse the status of land degradation.

China is one of the countries seriously affected by desertification with large desertification area, wide distribution and serious harm. The national desertification area reaches 2.622 million km^2, accounting for 27.3% of the total land area, covering 598 counties in 13 provinces (autonomous regions and municipalities) and near 400 million people are under the impact, with the annual direct economic loss amounting to over 54 billion yuan. At present, the pace of desertification further accelerates, causing rapid increase of dust weather, resulting in enormous economic loss and serious ecological consequence; the control speed can not keep up with the pace of destruction. Serious soil erosion in China is almost in all major river basins. The current soil erosion area reaches 3.67 million km^2 accounting for 38% of the total land area; the annual soil loss due to soil erosion amounts to 5 billion tons; the arable land destroyed by soil erosion reaches 2.7 million km^2 with the annual loss of about 60,000km^2; the national lake area reduces 1.86 million km^2 as a result of sedimentation, accounting for 40% of the existing lake area, and the lowering flood diversion capacity leads to frequent disasters. The soil erosion in China is shown as large erosion area, wide spreading range, high development speed, high erosion modulus, large quantity of sediment loss and serious ecological risk.

Desertification

The word "desertification" first appeared in a scientific paper of Lavauden in 1927. He used this word to describe the landscape in Sahara desert and pointed out the desertification in the region was entirely caused by human factors. In 1977, UN Conference on Desertification adopted the term of "desertification" and clearly defined it as, "the weakening and destruction of land biological breeding potential finally leads to the desert like situation; it is an aspect of generally deteriorating ecosystem, which weakens or destructs the biological potential". The nature of desertification is land degradation, loss of land resources and the appearance of desert like landscape. In 1992, United Nations Conference on Environment and Development (UNCED) briefly defined desertification as: as a result of the factors like climatic variations and human activities, the land degradation occurs in arid, semi-arid and sub-humid arid areas. The definition has also been adopted by the Convention to Combat Desertification.

Desertification is the product of the mutual-impact and interaction between strong human activities and fragile eco-environment, the result of contradictory man-land relationship. China is one of the countries seriously threatened by desertification. According to the Convention to Combat Desertification and the actual situation in China, desertification can be divided into the following main types, namely, water erosion, sandy desertification and salinization. In the past 50 years, the development speed, range and damage degree of those types were increasing gradually. The major form of desertification in most areas is sandy desertification.

Source: Zhu Zhenda. 1998. Concept of land desertification in China, causes and control. Quaternary Research, 5 (2): 145-155

Wang Tao, Zhu Zhenda. 2003. Study on sandy desertification in China–definition of sandy desertification and its connotation. Journal of Desert Research, 23 (3): 209–214

5.3 Serious Water Resources Shortage and Severe Water Environment Deterioration

The available fresh water on Earth surface only accounts for 0.014% of the total global water and is quite unevenly distributed. Because of rapid global population growth, the fast industrial development has resulted in a rapid increase in water demand of human being, and the modern industrial civilization has brought unprecedented challenges to water resource environment, such as a great deal of pollution to water ecosystem, species reduction, fish extinction, and serious degradation of the ecological system around the waters. In addition, the degraded environment of water resources further leads to water distribution imbalance, serious flood disaster, river cutout and aggravated soil erosion, which have polluted and eroded a great deal of water resources for consumption. In 20th century, the world's population has increased by 2 times, while the water consumption has increased by 5 times. According to the statistics of the United Nations, about 1.1 billion people have no access to clean water every day; and about 6,000 children die from unsafe water and the diseases caused by poor sanitation everyday. Along with the exacerbated hazard caused by increased water pollution, "environmental refugees" continue to increase; many countries in the world are facing water crisis. It is estimated that by 2025, the population have no access to safe water will increase to 2.3 billion, and there will be 3 billion people living in water shortage and regular water-scarce countries. At the same time, the deterioration of the ecosystem and biodiversity destruction brought by water crisis will also be a serious threat to human survival.

China's fresh water resources only account for 8% of the world and the per capita water resource amount is only 1/4 of the world's average level, thus China is an internationally-acknowledged water-lacking country. The water resources in China have the following four characteristics. The first one is serious water shortage. Beyond the serious urban and industrial water shortage, the national grain output reduction caused by water shortage is about 75 billion to 100 billion kg; about 80 million rural population and more than 40 million livestocks are lack of drinking water. The second characteristic is the low efficiency of water utilization and serious waste of water resources. Currently, the utilization coefficient of China's agricultural irrigation water is only 0.3 to 0.4 and the agricultural production efficiency on water is 0.8kg/m^3, less than half of that in the developed countries. The third characteristic is unreasonable exploitation, coupled with the lack of scientific planning and co-ordinate scheduling of upstream and downstream river water use, thus the exploitation of groundwater has given rise to a series of ecological degradation issues. The fourth characteristic is no effective control of water pollution. The untreated sewage discharged from cities, mines and enterprises has polluted 20% to 30% parts in the seven major rivers and five major lakes; the fourth grade pollution section in Yellow River Basin only has reached 12,000km, accounted for more than 60% of the total length of main stream and branches.

Serious Situation of Water Resources in China

In an interview at the beginning of 2006, Mr. Wang Shucheng, the minister of Ministry of Water Resources clearly pointed out that China's water conservancy had a quite arduous task and faced a serious situation. The four major water resource issues, namely, drought and water shortage, floods, water pollution and soil erosion were still prominent. Large population and water scarcity, uneven spatial and temporal distribution of water resources as well as the un-matching water resources and socio-economic development constitute the basic water situation in China.

The long-term extensive economic growth mode exacerbated the severity of water resources. Now we are faced with the following four tough issues.

Firstly, the prominent contradiction between water shortage and the growing demand in water resource of socio-economic development. The per capita water resource of China is 2,200m^3, about 1/4 of the world's average level, and the water shortage is near 40 billion m^3 in normal years. The development and utilization degree of water resources in some basin and region has been close to or exceeded the carrying capacity of water resource and water environment. Along with the socio-economic development and the improvement of people's living standards, the demand of water resources is in growing trend, while the difficulty of water resources development and utilization as well as river management keeps on increasing, thus the problem of water shortage will continue exacerbating.

Secondly, the prominent contradiction between the weak comprehensive flood control system and the protection of people's life and property. Most important tributaries of major rivers as well as the small and medium rivers have not been effectively managed, with backward construction of flood diversion area as well as low monitoring and defense capacity of disasters such as mountain torrent and debris flow. Each year, a large number of casualties and property losses are resulted from floods, which is still a serious hidden trouble of Chinese nation.

Thirdly, the prominent contradiction between serious water pollution and soil erosion and sustainable development. The national sewage discharge amount keeps on rising year by year, pollution in rivers and lakes is in exacerbating tendency and the overall situation of water environment has not been fundamentally improved. The trend of soil erosion and ecological deterioration has not been fundamentally curbed, the groundwater has been seriously over-exploited, and in some regions there have the phenomena of river cutout, lake drying up, wetland shrinking, and oasis disappearance, which gravely affected the sustainable economic and social development.

Fourthly, the prominent contradiction between backward rural water conservation development and the building of new socialist countryside. The rural water infrastructure is weak, 1/3 of national irrigation area is low-yielding field, many backbone buildings in large-scale irrigation areas have been damaged and the aging rate of large drainage pumping stations is high. Agricultural production still not gets rid of the passive situation of "heaven disposal". In rural area High fluoride water, high arsenic water, brackish water, etc., still gravely threat the health of farmers.

China's total water consumption is above 550 billion m^3. According to the current normal needs and non-exploitation of groundwater, the annual water shortage is near 40 billion m^3, more than 400 of the total 660 cities nationwide will be in water shortage, and 110 cities will be severely short of water. Hundreds of millions of farmers do not have clean water to drink, over 23 million people and 13 million livestock have temporary water drinking difficulties.

This is the serious situation of water resources in China.

Source: http://www.ce.cn/cysc/cysczh/200602/191t20060219_6130805.shtmt

5.4 Tightening Bottleneck on Environment and Resources, Exacerbating and Spreading Trend of Environmental Pollution

Currently, the most prominent contradictions of China's environmental issues are as the follows: firstly, during the process of industrialization, the development of the industries such as paper making, brewing, building materials and metallurgy, etc., has resulted in exacerbating environmental pollution and ecological destruction; secondly, the coal-dominated energy structure will exist for a long time, thus the treatment task of sulfur dioxide, soot, dust, etc., will become more difficult; thirdly, the construction of infrastructure during urbanization process lags behind, with the problems such as garbage and sewage can not be properly handled; fourthly, in the agricultural and rural development process, the use of chemical fertilizers and pesticides, as well as disorderly development of aquaculture, have aggravated rural environmental pollution, not only threatening the health of farmers, but also posing a threat to the safety of agricultural products; fifthly, during the social consumption transition, the new types of pollution such as electronic waste, vehicle exhaust, toxic building materials and interior decoration misconducts are in rapid increasing trend; sixthly, the new technology and new products such as genetically modified products and new chemicals will bring potential threat to environment and people's health.

According to the relevant research achievements of Chinese Academy of Social Sciences, in late 1990s, China's economic loss caused by various types of environmental pollution was as high as 180 billion yuan annually, among which the loss resulted from water pollution was 144 billion yuan. The results of Study Report 2004 for Green National Economic Accounting show that the economic loss caused by environmental pollution in 2004 was 511.8 billion yuan, accounting for 3.05% of the GDP in that year. In 2008, the Yale center for Environmental Law and Policy released the first official Environmental Performance Index (EPI) at the World Economic Forum in Davos, which quantified and numerically scaled the environmental performance of 149 countries and regions. The score of China was 65 and ranked No. 105.

5.5 Increasingly Prominent Urban Environmental Issues with Structured, Composite and Compressing Characteristics

The next 30 to 50 years will be an important period for the rapid urbanization process in China. Along with the rapid urbanization process and urban agglomeration development, there appears such tough situations as water shortage and over-exploitation of water resources, inefficient energy use, space development and land use being out of control, traffic congestion, and increased waste discharges, coupled with the overall backward construction of urban environment, which has exacerbated urban heat island effect, crowding effect and environmental pollution, not only leading to the deteriorating surviving and living environment of the urban residents, but also bringing serious impact on the ecosystem and environmental quality; it has become a major bottleneck and challenge faced by China's urbanization process and socio-economic sustainable development.

Along with industrialization and the transfer of rural surplus labor force, the urbanization level of China will be rapidly enhanced in the coming decades, which will exceed 50% in 2020; at the same time, urban pollutant discharge will significantly increased, and the national urban domestic sewage and garbage will increase by 1.3 times and 2 times respectively, compared with that in 2000. By 2030, China's industrial water demand will increase by over 5 times. If effective measures not be taken, a great deal of urban waste, electronic waste, abandoned vehicles, industrial hazardous waste, new chemical substances and other hazardous wastes will enter into the environment, which will enhance the difficulty to urban ecological protection and pollution control and derive a serious secondary pollution. In addition, the backward industries in big cities will be transferred to the rapid developing small and medium size cities and towns, thus further aggravating environmental pollution. With the rapid increasing vehicle number, if it fails to timely increase the vehicle exhaust emission standards and fuel quality, urban vehicle emissions in 2010 will be over doubled than that in 2000.

The highly intensive urban development and rural urbanization will increase the pressure to water environment and ecosystem, resulting in the obvious change of the urban natural water system, vegetation pattern and species composition, the reducing area of farmland and protected field as well as lowering regional ecosystem regulation capacity. Rural environment and urban environment are inseparable. The urban pollutants, including water, waste gas and solid waste, generate huge impact on rural areas; on the contrary, all kinds of pollutants generated in rural area and agriculture are and will be more significantly impacting the urban environment.

According to the statistics of the World Bank, among the 20 most polluted cities in the world, there are 16 in China. The current issue of the urban environment shows the characteristics of structure type, complex type and compression. The environmental problems occurred in phases during the century-long industrialization process of developed countries collectively took place in recent 30 years, the most rapidly developing period in China, such as large pollutant discharging amount and increasing complexity of industrial structure. In recent years, environmental crises caused by unexpected urban environmental pollution incidents occurred frequently (such as the Songhua River and Taihu Lake water pollution incidents, etc.); only during the period from January to middle May in 2006, the State Environmental Protection Administration received 49 reports of unexpected environmental incidents, involving 22 provinces (autonomous regions and municipalities), showing that China's current urban environmental issues have generated a series of shocking punitive impacts on socio-economic development. It can be said that, compared with traditional environmental pollution in foreign countries, China's modern urban environmental issues have significant changes in the aspects of causes, characteristics and hazards, with the hidden hazard of collective outbreak at anytime, which has become a huge barrier to the socio-economic sustainable development.

5.6 Gradually Emerging Hazards of New Pollutants and Persistent Organic Pollutants Lead to Increasing Ecological and Environmental Risks

Some new pollutants, such as antibiotics, endocrine disruptor, algae toxin and oxidation by-products of pesticide have greater risks as well as long-term and unpredictable potential impacts on ecosystems, food security, and human health, etc; the hazard of persistent organic pollutants aggravates. Toxic pollutants in the environment are not only toxic, but have many other effects, including carcinogenicity, teratogenicity, mutagenicity, also with endocrine disrupting effect, directly threatening survival and reproduction of wild animals and even human beings. A number of toxic pollutants (such as persistent organic pollutants) are hard to be degraded in environment (long residence time) and with high-fat solubility (low water solubility), which can gather in food chain and affect the regional and global environment through evaporation-condensation, air and water transmission.

Presently, targeting at the hazards of toxic pollutants, the developed countries and organizations have launched many major programs and carried out the monitoring, screening and management on toxic pollutants, such as the Registration, Evaluation, Authorization and Restriction of Chemicals (REACH) made by EU, strictly controlling pollutants entering environment and proposing toxic and ecological management requirements towards the management of toxic pollutants.

The "dose – effect" relationship, used to be the evaluation criteria of health risk, can not explain the long-period and low-dose exposure toxic effects of persistent toxic chemical pollutants. Some of the new observations and new findings mentioned above suggest that there are great deficiencies in the past assessment approach system of health risk, which is in negligence of the role of low-dose exposure and the low-exposure level of people living in environment surrounded by hundreds of synthetic chemicals, thus the existing chemical security assessment system must be re-examined and adjusted. The existing environmental chemistry and eco-toxicology theory of chemical pollutants have a lot of limitations, whose theoretical system is brewing a great innovation.

Persistent Organic Pollutants

Persistent organic pollutants (POPs), refer to the natural or synthetic organic pollutants with the characteristics of long-term residue, bioaccumulation, semi-volatilization and high toxicity, which can migrate in atmosphere with long distance and deposit to the earth, which pose a serious threat to human health and environment. POPs has 4 distinct characteristics:

(1) Persistence / long-term residue. POPs have a strong resistance to biological metabolism, photodegradation, and chemical decomposition, etc, under natural conditions. Once released into the environment, they are difficult to be broken down; therefore, they can exist in environmental media such as water, soil and sediment for several years or even decades or longer.

(2) Bioaccumulation. The molecular structure of POPs usually contains halogen atoms, with low water-soluble and high-fat-soluble characteristics, thus bioaccumulation can occur in adipose tissue, causing the concentration of POPs in biological body from the surrounding media and reaching toxic concentration through the food chain bio-magnification.

(3) Semi-volatilization. POPs can enter the atmospheric environment or be adsorbed on particulate matters in the form of vapor from water or soil, and have long-distance migration in atmospheric environment; at the same time, the moderate volatilization will not ensure them a permanent stay in the atmosphere, but to resettle on earth and this process can occur repeatedly.

(4) High-toxicity. Most POPs are highly toxic to human beings and animals. The laboratory studies and epidemiological surveys have shown that POPs can lead to serious illnesses such as endocrine disorders, reproductive and immune dysfunction, neurobehavioral and developmental disorders and cancer.

The half-life of POPs in water is mostly from dozens of days to 20 years, some can reach 100 years; and that in soil is mostly from 1 to 12 years, some can reach 600 years; while that of their bio-concentration factor (BCF) is as long as 4,000 to 70,000. The above mentioned characteristics determine the great harm of POPs on human health and ecological environment.

Source: Yu Gang, Huang Jun, Zhang Pengyi. 2001. Persistent organic pollutants: one of the important global environmental problems. Environmental Protection, 4:37–39

6 The Development Trend of International Ecological and Environmental Science & Technology

6.1 Global Climate Change Mitigation and Adaptation Have Become the Focus of Earth System Science

Earth system is a complex non-linear multi-coupling system consisting of a series of interaction processes, including the interaction among earth system layers, interaction among physical, chemical and biological processes as well as the interaction between human and earth environment. The whole view of earth system, research on the interaction among the three basic processes, and the special concerns on the impact of human activities on earth environment have made the global change science a new integrated science in the forefront of contemporary international science.

The new research framework of global change science focuses on the three major sub-systems of earth system, namely, atmosphere, ocean and land, as well as the interface among them. Concerning on the issue of global sustainability, the Earth System Science Partnership (ESSP), that is, DIVERSITAS, the International Geosphere – Biosphere Programme (IGBP), International Human Dimensions Programme on Global Environmental Change (IHDP) and the World Climate Research Programme (WCRP) co-sponsored three joint projects: global carbon cycle system, water resources system, global environmental change and food system. How to better identify the response of climate change in monsoon area to global change and their association, how to better forecast future climate change, how to deal with climate change and mitigate the adverse impact of climate change on China's economic and social development, and how to formulate and implement national strategies to cope with climate change have become the focuses of international earth system science research.

Since the 21st century, climate change has been exacerbating, the impact of human activities has been increasing significant and the issue of global warming has aroused wide concern in the international community. In order to scientifically assess the research progress of climate change, particularly the fact of climate change and its impact, IPCC, co-established by World Meteorological Organization and United Nations Environment Program, launched the fourth assessment report in 2007, clearly stated that human activities had exerted an overt impact on global climate, especially in the past 50 years. Moreover, on the basis of global surface temperature rise (0.74±0.18)°C from 1906 to 2005, the report estimated that at the end of 21st century, global climate will continue to rise 1.1°C to 6.4°C. In order to cope with the occurred and occurring global warming as well as a series of impacts, governments and international organizations have launched large-scale scientific integrated research programs of climate change or emphasized the importance of climate change research in scientific planning, for example, in the scientific planning of NASA from 2005 to 2020 and National Center for Atmospheric Research (NCAR) from 2005 to 2015, and the earth system observation and prediction coordination research plan from 2005 to 2015 proposed by ICSU, World Meteorological Organization and International Oceanographic Commission, etc. Global climate change has been put in priority.

6.2 Relationship between Ecosystem and Human Well-being Has Become the Main Object of International Ecological Research

The "Millennium Ecosystem Assessment (MA)" launched by the United Nations in June 2001 is a four-year international cooperation project, aiming at providing the decision makers with scientific information between ecosystem and human well-being. *Ecosystems and Human Well-being: A Framework for Assessment* is the first achievement of the project. The main assessment report was published in 2005. MA work mainly focuses on the following aspects: how does the change of ecosystem service function affect human well-being? In the coming decades, what impacts will be brought by the changes in ecosystem? What kinds of countermeasures should be taken by human beings on regional, national and global scale to improve ecosystem management, thus enhance human well-being and eradicate poverty?

Ecosystem assessment can provide assistance for all countries in the following aspects: to have a deeper understanding of the relationship between ecosystems and human well-being; to demonstrate the potential of ecosystem in eradicating poverty and improving human well-being; to assess the compatibility of policies formulated by agencies at all levels; to integrate the economic, environmental, social and cultural objectives; to integrate the information of natural and social sciences; to confirm and evaluate the sustainability for ensuring ecosystem service function as well as the needs of ecosystem and human being; to coordinate the formulated policies and management countermeasures; and to promote comprehensive management of ecosystem.

In face of the degrading ecosystem, human beings have a growing demand of its service functions, which seriously impacts the prospect of sustainable development. Human well-being not only influenced by the supply-demand gap of ecosystem service function, but also subjected to the rising vulnerability at individual, community and national level. Ecosystem, with high productivity and its service function, can provide important resources and countermeasures for human beings and human community to cope with natural disasters and social upheaval thus protect human security. Ecosystem at high management level is conductive to reduce risk and vulnerability; while poor managed ecosystem may increase the disaster occurrence rate, like flood, drought, poor harvest and diseases, hence further aggravate the risk and vulnerability.

Because of the inertia characteristics of ecosystem and human system, the current changes in the ecosystem may be imperceptible in the next few decades. Therefore, to obtain a sustainable ecosystem service and human well-being, efforts must be made to gain a comprehensive understanding and reasonable management on human activities and changes in ecosystem, and also the relationship with human well-being on short, medium and long-term time scales. The present excessive utilization of ecosystem service function is bound to endanger its future prospect, while such crisis can be avoided through sustainable utilization methods. In order to deal with the scientific and technological need, ecologists pay growing concerns on how to solve the contradiction between the growing need of human beings towards ecosystem product and service functions (like food and clean water) and the lowering ecosystem capacity caused by human activities through ecosystem management.

6.3 Integrated Research of Ecosystem and Human Activities, Natural System and Economic System Has Become the Main Idea in the Study on Global Ecological Issues

The solution of global ecological issues involves quite a few aspects such as characteristics of global geographical distribution, resource and environment of natural system, social economic system and impact of human activities, etc., how to conduct comprehensive integration on the basis of single mechanism will become the key of research. The single institute and single discipline research approach of the single resource, environment or disaster issue in the past has been replaced by the research mode of interdisciplinary, multi-sectoral approach that targeting at the single or comprehensive resource, environment or disaster issue made by natural scientists, social scientists, economists, engineers and policy makers. No matter in international scientific community or national research organizations of countries in the world, the basic research of the major resource, environment and disaster issues is usually completed through setting special research programs.

After entering into the 21st century, ESSP, co-sponsored by the four major research plans of global changes, namely, DIVERSITAS, IGBP, IHDP and WCRP, takes the whole earth system as the research object. The cross-disciplinary integration research on different fields such as environment and development, nature and humanities, technology and policies, etc, is increasingly clear and further, not only promoting the development of traditional disciplines, but also generating many new disciplines.

Practice has proved that the research on complex earth process and its major resource, environmental and ecological problems can not be completed by a single discipline as well as traditional concepts and approaches. Many breakthrough opportunities, new growth points and solutions are contained in many important crossings, which demands interdisciplinary cross, penetration and comprehensive integration. At the same time, the relationship between man and nature puts forward growing demand for the crossing of natural science and social science. The integrated study of ecosystem and human activities, nature system and economic system has become the main idea of addressing global ecological issue. At present, the solution of global ecological issue has crossed the field of ecology, which is not only the interdisciplinary study of biology, earth science, forestry, agriculture, soil science and meteorology, but emphasizes the coupling of social subjects and humanities. This is because the solution of global ecological issue depends on global effort, not only the natural process, but needs the support of the whole society, including political, economic and human intervention. Its effectiveness depends on political support and the assurance of legislation and relevant institutions, emphasizing the full cooperation among scientific workers, government and people, thus solving global ecological issue needs the integration of ecology, economics, sociology, law and political science. These will inevitably promote the convergence, crossing and infiltration among neighboring disciplines. The discipline convergence, theory and method transplantation enhance the level of ecological discipline and constantly explore new research areas, thus forming new edge disciplines and inter-disciplines.

6.4 Quantitative Assessment and Scientific Prediction of Ecological and Environmental Changes Have Become the Scientific and Technological Objective of Ecological Research

The core issue of global change science and ecological science in 21st century is the ecosystem change co-driven by highly intensified human activities and rapidly changing global environment. In response to this core issue, the scientists use multi-scale joint observation of ground-based ecological networks and satellite remote sensing, cross-scale mechanism analysis and simulation prediction to recognize the environmental as well as human-driven mechanism of ecosystem changes through the combined application of multidisciplinary approaches. The scientists also conduct quantitative estimation and prediction of the changes in regional ecosystem patterns and key process under the background of global environmental change to recognize the impact of human activities and environmental changes on ecosystem service functions and sustainable development capacity. Especially the establishment and application of earth observation system, GPS and satellite network communication system have greatly improved the efficiency and quality of ecological research. Currently, a comprehensive study of large region and analysis of ecosystem structure, function and evolution of time and space can be carried out. Along with the experience accumulation and gradually improved method, the quantitative method will be generally applied and gradually become precise. Mathematical models can express the experience concept and experimental results and also reveal the internal relations and the law. The combination of computer and space technology makes geographic information system become a powerful technological means of modern ecology. With the development of geographic information system towards the aspects of uniform standard, multi-level distributed system, close connection with remote sensing technology as well as the further development of intelligent analysis and decision-making at expert system level, etc., the comprehensive research of ecosystem will definitely have a new breakthrough.

Make assessment of main ecosystem types and major ecological issues in different areas, including the four interrelated parts, namely, the restoration of the past evolutionary history, assessment of present status, future change prediction and countermeasures of ecosystem optimization management, is a key approach of ecology to provide scientific basis for management and decision-making. Through quantitative monitoring, simulation, evaluation and analysis on changes in regional natural environment and ecological environment, we could scientifically predict various statuses of ecological environment generated by different regional developing patterns in the future to provide basis for adjusting and optimizing the overall and partial regional ecosystem management.

6.5 Network Observation of Ecosystem, Simulation Experiment and Virtual Numerical Simulation Serve as the Primary Means for the Study on Comprehensive Ecological Issues

The study on global environmental issue and development of global change science require natural and social science circles to provide more detailed and reliable data and information of global changes, in particular the effective information of the impact and function of human factors. As an important means of studying the relationship between global change and terrestrial ecosystem, the network observation and research of terrestrial ecosystem can combine with the satellite remote sensing, geographic information system technology and mathematical model to analyze and simulate the ecosystem patterns and process changes on landscape, regional and global scale. Therefore, a number of regional ecosystem observation and research networks have been established by the international community, respectively around the process of global climate change, biogeochemical processes, socio-economic process, etc, to carry out international and cross-regional cooperation in joint observation, network experiment and comprehensive study.

Due to the change of research object and task, ecological research gradually develops towards regionalization and globalization on the basis of relatively isolated local area and forms network for comprehensive and comparative study. The long-term ecosystem observation and research has experienced the development course from single station research to network study and integrated research, and become a key platform of global change research, with the research direction changes from concerning on classic ecological issues to the combination of natural science and humanities, adopting the model of network development driven by long-term ecological research project plan. Particularly in the study on the relationship between global change and terrestrial ecosystem, the transect-based networking process has become an important research platform and effective way. For example, MAB program has established a biosphere reserve network composed of 261 positioning stations in global context, IGBP has planned and established 17 terrestrial transects, including the northeast China transect (NECT) established in 1993 and north-south transect of eastern China (NSTEC) established in 2000, as well as Inter-American Institute for Global Change Research (IAI), the Asia-Pacific Network for Global Change Research (APN) and the European Network for Research in Global Change (EN-RICH) and so on. Meanwhile, some countries have established or are establishing their own ecological research networks, such as the Chinese Ecosystem Research Network (CERN), the US Long Term Ecological Research (LTER) Network, the Environmental Change Network (ECN) of United Kingdom, Canadian Ecological Monitoring and Assessment Network (EMAN), etc. The main research contents of those research networks include: ecosystem regulation technology for representive regions or ecological types, the impact of environmental change on ecosystem succession, optimization management, the impact of human production activities on key ecological processes and the regulation technology.

Establish ecosystem observation network based on the complexity and long-term nature of ecological and environmental issues, covering long-term observation and testing of ecosystems in different regions and of different types, conduct modeling, numerical method, software and basic research based on numerical simulation in scientific and engineering calculations by combining with a number of major scientific issues in complex system, adopt simulation experiments, remote sensing, model simulation and other technical means, as well as the technical means such as multi-scale observation, multi-method proof, multi-process integration, and cross-scale simulation to recognize the changing laws of ecosystem structure and function, process and patterns, which are jointly-driven by global change and human activities, and explore the means of control management in the structure, pattern and process of ecosystem, thus become major means to address food production, natural resource management, ecological environment protection and cope with global changes, and so on. On the basis of regular meteorological observations, WCRP has carried out continuous observation and monitoring on a large number of environmental factors and the background value by means of ground fixed points, mobile and satellite observation, remote sensing and telemetering, etc; IGBP has implemented global carbon flux observation scheme and DIVERSITAS has conducted fixed-point observation and transect survey by relying on global ecological experiment station.

6.6 Environmental Technology / Green Technology Based on Material Cycle Has Become a Fundamental Approach to Solve Environmental Pollution

In the 21st century, the speed of knowledge dissemination and update will further accelerate; the transformation and application pace of technical breakthroughs in the areas such as material, life and information is accelerating in manufacturing sector; the deep integration of high-tech and manufacturing promotes high technicalization of traditional process industries. This will facilitate significant changes in production techniques and ways of process industry; industrial ecology and resources recycling has become the development direction for the process industries in developed countries to improve the efficiency of resource utilization. Seen from the domestic and international technology development, clean production and circular economic technology realize energy conservation and source pollution abatement through green process and technology, achieve construction of industrial ecosystem through recycling resource utilization, with the ultimate goal of realizing a high degree of harmonious co-existence of production process, consumption and natural ecosystem. Countries in the world, especially the developed countries are focusing on the development of clean production and circular economy; the essence is to reduce resource consumption and pollution emission while keeping substantial increase in economic growth and social development, so as to achieve the decoupling of pollution control from economic development. Germany and Japan are in the leading position of developing circular economy; however, their development of circular economy is located in the environmental management model and emphasizes solid waste recycling. In pollution control and environmental-oriented management, the environmental technology / green technology based on material cycle will become the core in future development. This mainly includes three aspects: firstly, environmental pollution assessment and control technology of the use of waste resources; secondly, the highly efficient and clean resource transformation utilization technology; thirdly, enterprise network and eco-industrial integrated system; it is necessary to develop ecological industry and ecological agriculture technology, and explore the building model of eco-city (county). Through the research and development of those technologies, conduct resource utilization and decontamination of the pollutants generated in people's production Processes and living during the whole material recycling system to reduce the pressure on environment, which is the fundamental approach to the issue of environmental pollution.

6.7 Concurrent Consideration of Ecosystem and Human Health Has Become the Theme of Environmental Health Research

In recent years, ecologists began to explore human health problems in ecological point of view. The viewpoint of ecosystem health expands our understanding of human health as well as the characteristics and source of human diseases. Many of the global epidemic diseases (especially those spread by insects) are quite sensitive towards the climate and weather, while little is known about the causal relationship between environmental factors and epidemic disease activities. In recent decades, new diseases increased at an unprecedented rate, and often spread across borders to the world; at the same time, traditional diseases such as flu pandemic, malaria and tuberculosis continue to produce varieties with increasing drug resistance. Those not only relate to the immune system status of human bodies and human health care system, but also closely relate to the health status of ecosystem. In recent years, the World Health Organization, United Nations Environment Program and its partners as well as the member states were in close cooperation, made joint effort to promote the formulation of environmental and health strategies and policies, proposed a series of suggestions for strengthening environment and health work, emphasized the establishment of long-term inter-institutional cooperation mechanism between environment and health sectors and worked out national environment and health action plan to promote positive development of environment and health work.

Human activities will threaten ecosystem health and result in changes in ecosystem, thereby affecting the service functions of ecosystem and having an impact on human health, and then human beings have to pay concerns on the issue of ecosystem health. The current research on population health has transformed from traditional disease control to the comprehensive exploration of welfare, quality of life, human physical state, the relationship between psychological state and ecosystem health. From the point of view of research approaches, the integration and comprehensive application of traditional method of epidemiological investigation, medical geography research method, geographic information system and spatial analysis, environmental toxicology research method, eco-environmental planning and management, etc, will be conductive to addressing the issue. The study on environmental health will take both ecosystem and human health into consideration to explore an integrated approach for human health improvement.

6.8 Multi-scale Environmental Biogeochemical Process, Mechanism as well as the Coupling Characteristics and Modeling Are Important Means to Study Regional Environmental Issues

The biogeochemical process of human system is regarded as one of the five main driving factors that lead to global environmental change. A variety of waste materials generated by highly intensified human activities enter into the atmosphere, water and soil through the migration of their own or carriers, thus affecting the biogeochemical cycle from the city scale to global scale. Biogeochemical cycles can occur in the process from the molecular level to the global spatial scale. From the viewpoint of system ecology, the process of small-scale has a significant impact on the fluctuation of macro-process, while large-scale study can provide a higher stability to research results. It can make up for the patterns and processes beyond the observation of single small-scale or large-scale study by means of scaling. Under the support of U.S. National Science Foundation, the U.S. ecologists are preparing for the establishment of NEON, which will support the continental-scale ecological prediction through providing data and other equipments to address the above-mentioned issues. The needed data scope is from genome to continental scale with the time ranging from seconds to decades. The comprehensive act of countless organisms will control the atmosphere transfer and chemical characteristics, regulate the surface substances on the earth and influence the supply and quality of water. Organism scale and different scales of its role in global environment have given rise to the major issues of large-scale ecological research. Natural life is the result of the longest time and the largest spatial scale, and the biosphere process must be known through the record of organism, ecological communities, biomass, and other small-scale phenomena.

The spatial heterogeneity of soil system, drainage basin and offing increases under the strong impact of human activities. The intertwined natural and humanistic material cycles resulted in more complex biogeochemical processes on different scales. The current global and regional environmental pollution shows highly complex characteristics: ① complex attribute of spatial representation, covering a wide range of ecosystem types from land, drainage basin to offing; ② complex medium attribute, with multi-phase pollution superposition of water, soil and air from surface to underground; ③ complex attribute of formation pressure, with interacting coexistence of drainage basin socio-economic types and development stages; ④ complexity attribute of pollution source, with complicated pollution sources of industrial and agricultural production as well as human inhabitancy, and a variety of pollutants; ⑤ complex attribute of pollution process, with chemical pollution, eutrophication, microbial pollution and toxic organic pollution in various processes; ⑥ complex attribute of pollution effect representation, with different damaging characteristics of ecological integrity in different basins, having a significant risk mechanism difference on human and ecosystem health.

Multi-scale environmental biogeochemical process and mechanism as well as the coupling characteristics and modeling are the keys for identifying system dynamic evolution. Under the background of global environmental change, ascertaining multi-scale environmental and biogeochemical process, mechanism as well as the coupling characteristics and modeling are important research means to study environmental issues.

7 Key Research Areas of Ecological and Environmental Science & Technology

In *Outlined Plan for China's Mid- to long-term Science and Technology Development (2006–2020)* published in 2006, environment was listed as the third among the 11 key areas, manifesting the state's emphasis on ecological protection and environmental construction. There are 4 prior themes in environmental area, including:

(1) Comprehensive pollution control and waste recycling. Mainly focus on developing monitoring and early warning technology of regional environmental quality; make breakthrough on key technologies such as air pollution control of urban agglomeration and so on; develope non-conventional pollutant control technology, waste resource utilization technology, and the clean production integrated technology of heavy pollution industries; establish and develope the technological demonstration mode of circular economy.

(2) Restoration and reconstruction of ecosystem function in ecologically fragile regions. Focus on developing a series of technologies, including developing dynamic monitoring technology, grassland degradation and rodent control technology, restoration and reconstruction technology of degraded ecosystem in the typical ecologically fragile regions, such as karst region, the Qinghai-Tibet Plateau, the middle and upper reaches of Yangtze River and Yellow River, the Loess Plateau, desert and desertification areas, agriculture and animal husbandry ecotone and mineral mining area, developing ecological protection and restoration technology around the major projects like the Three Gorges Project, the Qinghai-Tibet railway line and the complex mining area. Establish the technical supporting models of ecosystem function restoration and continuous improvement of different types of ecosystems, and build a comprehensive and technical assessment system of ecosystem.

(3) Protection of marine ecology and environment. Focus on developing marine ecological and environmental monitoring technology and equipment; strengthening technological research on marine ecological and environmental protection; developing coastal water ecological and environmental protection and restoration, marine emergency treatment technology; and developing high-

precision ocean dynamic environment numerical forecasting technique.

(4) Monitoring and countermeasures of global environmental change. Focus on researching and developing accurate monitoring technology of large-scale environmental change; developing the emission control, deposition and utilization technology of greenhouse gas, such as carbon dioxide and methane in main industries; developing bio-carbon fixation technology and carbon fixation engineering technology; and carrying out countermeasure research on climate change, biodiversity protection, ozone layer protection and persistent organic pollutants control.

China issued National Ecological Construction Planning in 1999, which planned the ecological environment construction would be implemented at short, medium and long-term phases before 2050, and proposed an overall objective: using 50 years' time, mobilize and organize people all over the country through relying on science and technology to strengthen the existing natural forest and wildlife resources, to plant more trees and grass, to control soil erosion and desertification, to build ecological agriculture, to improve production and living conditions, to strengthen comprehensive management and complete a batch of projects which are conductive to improve national ecological environment, and to reverse the trend of ecological deterioration. Following the trend, in the mid-21st century, the national soil erosion area that is suitable for control will be basically renovated, the land suitable for greening will be planted with trees and grass, the grassland with degradation, salinization and desertification will be basically restored, a better monitoring and protecting system for ecological environment will be established, the ecological environment in most areas will be significantly improved, and nationwide charming landscapes will be basically realized.

Combined with the existing national science and technology development planning, based on the above-mentioned technological needs of China's socio-economic development and technological development to 2050 as well as the frontier of international ecological and environmental technology development, we propose the following eight key research areas of ecological and environmental science and technology.

7.1 Global Climate Change and Its Ecological Processes

The Fourth Assessment Report of IPCC manifested if exclude the influence of atmospheric aerosols and other climate forcing factors, only take the changes in atmospheric greenhouse gas concentration into consideration, the global warming would be even intensified in the past over half century; global warming in the past 50 years was most probably (>90%) attributed to the force of greenhouse gases; evidence shows that human activities have affected quite a few climate system components, such as global snow and ice reduction, water cycle acceleration, sea level rise and changes in vegetation. In a word, the climate warming caused by human activities is widely spread and the relevant signals have been detected in the surface of the earth, tropospheric, marine-related detections. The research objective of global climate change and its ecology process area is: to reveal the fact and the inherent dynamic mechanism of China's climate change on difference time scales; to scientifically evaluate the role of human activities in climate change of Modern China; to establish China's own model of earth system; to establish short-term and long-term forecast system of China's climate change on a variety of time scales; to uncover interaction among the global climate change, human activities and ecosystem process; to establish overall strategy and measures of China's adaptation to climate change in different areas; and to work out scientific strategies of China's international cooperation and diplomatic negotiation in climate change. The key scientific and technological issues of this area are as follows.

Climate Model

The main approach for estimating climate change caused by increasing atmospheric greenhouse gases (such as CO_2, CH_4, N_2O) is physical method. The so-called physical method is to determine the characteristics of various components of climate system based on the basic laws of physics; those laws are usually expressed by mathematical equations, which constitute the climate mathematical model, that is, climate model. Climate model can not only simulate contemporary climate, and can also be applied in simulating the climate change caused by some external condition changes. Therefore, if human beings are able to research climate and its changes through employing experimental methods, then the most important experimental equipment is climate model. Of course, any model is just a certain similarity to the actual climate system; at the same time of being employed in climate simulation, its own reliability degree must be continuously tested and improved.

Source: Zeng Qingcun, Zhang Xuehong, Yuan Chongguang. 1989. The concept, approach and status-quo of climate model. Advances in Earth Science, (3): 1–26

Climate System

The concept of climate system is one of the important symbols manifesting a new stage of climate research. In this sense, people not only study the impact of inner atmospheric process on climate change, but also need to consider the role of ocean, ice and snow, earth surface and biological conditions in climate change. That is to say, regard climate change as the overall behavior of the climate system composed of atmosphere, ocean, cryosphere, land surface and biosphere. The interaction of various physical, chemical and biological processes of the above-mentioned sub-systems determines the long-term average state of climate and various time scale changes.

Source: Li Chongyin. 1995. Introduction to Climate Dynamics. Beijing: China Meteorological Press

1. Climate Change of China in the Past and the Research on Its Mechanism

Under natural conditions, great climate changes occurred in the past; the research on the facts and causes has a great significance in the recognition of the inherent mechanism of climate change, the understanding of contemporary climate operation and estimation of the climate change trend in the context of global warming. Therefore, on one hand, we need to use a variety of proxy data to further rehabilitate and reconstruct the past climate changes in China in both typical warm period and cold period on the scales of earth orbit and structure; on the other hand, the paleoclimate numerical simulation research is needed to be strengthened to comprehensively assess the simulation capacity of China's climate model towards the past climate, promote the system research and development of climate model, and then make comparative analysis of the simulation results and reconstruction records to help the analysis and interpretation of ancient climate information, so as to ultimately research and reveal the internal mechanism of natural change in China's climate on a longer time scale from the perspective of climate dynamics.

The past few thousand years is an important period of converging geological records and measuring information; during this period, the impact of human activities on earth environment gradually increased and overlaid with the natural climate changes; therefore, understanding the global climate change in this period is crucial to recognize the role played by human activities and natural variability in modern climate changes. During the course of reconstructing climate changes in Chinese history by using proxy data, it is necessary to simultaneously emphasize the use of climate model and simplify earth system model to conduct continuous and transient climate simulation of typical period, focus on the causes of climate changes in Chinese history through comprehensive comparison with the reconstruction records, and assess the individual or joint role of effective solar radiation change, volcanic activity, atmospheric greenhouse gases, and aerosol concentration, etc., particularly the scientific assessment of comprehensive impact of human activities on the climate in Chinese history, which is crucial to understand the future trend of China's climate change under the background of global warming.

The greenhouse gases and aerosol emitted by human activities as well as the changes such as land-use since industrial revolution, especially in the past century, have an increasing impact on global climate system. In three aspects, namely, proxy information, measuring information and numerical simulation, respectively and comprehensively analyzing the spatial distribution characteristics of China's climate as a whole or in different regions on decadal time scale of the past century and discriminating the impact of natural factors and human activities on the climate of China in the past century is crucial for understanding the background of contemporary climate of China and the role of human activities in the formation of the contemporary climate. Scientific evaluation of the internal mechanism of Chinese decadal climate change in the past century is also of great importance in forecasting the trend of Chinese climate as a whole or in different regions over a period of time in the future.

The meteorological disasters in China are somewhat serious, most of which can be attributed to weather and climate variability and extreme events. Therefore, in the context of global warming, the changes in weather, climate and extreme events are in urgent need of comprehensive study. At the same time of analyzing the spatial and temporal change characteristics of weather, climate and extreme events over the past few decades by means of various measuring records and re-analysis information, we also need to strengthen the assessment and test of the probable internal mechanism of those changes in recent decades from the perspective of weather and climate models. Meanwhile, the short-term climate prediction on seasonal to annual scale also needs strengthening in terms of statistics and numerical simulation, which is essential to the prevention of meteorological disasters and their secondary disasters.

2. Research and Development of Earth System Model as well as the Assessment of China's Regional Climate Change and Impact

A unique feature of earth system science is we can not carry out real experiment on the whole earth system like such disciplines as chemistry, physics and biology; therefore, scientific researchers gradually applied the basic laws derived from the relevant natural disciplines since middle 1970s, to progressively realize the process modules of the major operation laws in climate system and its subsystems in computer and then develop into effective tools for researching climate system, its subsystems and their interaction through the method of numerical simulation. Currently, the earth system model or the so-called earth simulator has become the most important research tool in the field of global change. As a result of relatively late start, coupled with the dependence of earth system R&D model on multi-disciplinary cross-cooperation, although China has made some achievements in the area of earth system model, there is still a great gap compared with the R&D level in the major developed countries.

At present, the assessment of China's regional climate change and its impact, the study of the temporal and spatial climate change rules and mechanism in China and the surrounding areas as well as the estimation of the future climate change trend are the basic information data needed by the government in scientific policy-making, which is in urgent need of taking climate system model with high precision and high spatial and temporal resolution as a research tool. In view of the complexity of the climate system act, the establishment of climate system model must simultaneously consider the processes such as the biogeochemical processes of land and atmosphere, land and sea, sea and atmosphere, as well as the interaction with human activities; the refined, integrated and rapid development of the research decides the necessity of combination among different disciplines. At the same time when taking relevant scientific research institutions as the unit to continue in-depth study on various climate subsystems, at the present stage, it is necessary to draw on the experience of the existing earth system models in the world, focus on the cross-joint of scientific researchers in different academic fields to jointly establish and develop our own earth system model through effective collaboration. In the process, the practical application issues of earth system model need to be specially emphasized; cross-disciplinary scientific research team should be deployed to carry out interdisciplinary large-scale numerical simulation in various areas, targeting at the past, present and future global climate, especially the climate in China; from the perspective of climate dynamics, we need to reveal the internal operation mechanism of China's climate change on different time scales to promote the overall development of national earth science.

3. Short-term Climate Forecast, Decade to a Century-scale Climate Prediction

Along with the increasing atmospheric concentration of greenhouse gases, the qualitative conclusion of global climate warming and the continuity over a period of time has been widely recognized. In the context of global climate change, short-term climate forecast and long-term climate change prediction are of great importance to scientific disaster prevention and mitigation, national macro-economic construction as well as social sustainable and harmonious development. Hence, we need to implement large-scale basic scientific research from the perspective of mathematical statistics and dynamic numerical model and find out the key climate factors that influencing the climate change in China on various time scales to enhance the real-time forecast level of short-term climate on seasonal to annual scale, thus provide more reliable climate prediction information to the relevant social fields.

In synchronization with establishing and developing earth system model, we need to conduct large-scale numerical simulation experiment, targeting at global climate, especially the climate in China in the condition of a variety of greenhouse gases and aerosols emissions; at the same time of positively participate in international prediction of climate change, focus on the probable changes of East Asian monsoon climate in the coming decade to a century and carry out multi-situation prediction of the weather and climate trend in different areas by means of statistical downscaling and dynamic downscaling, etc, thus provide basis for the national scientific assessment of climate and environmental changes. During the period, we are also required to draw on the prediction results of the internationally advanced climate system model to deepen the understanding of future changes in China's climate through multi-modal superset prediction.

Global warming, together with the accelerated hydrosphere circulation will impact the status of global land-based glaciers and water resources to a great extent. China has a large population, with relatively poor ecological environment in the central and western regions and overall shortage of natural resources, thus the social sustainable development is under immense pressure. With intensified climate change and economic development, the pressure of scarce water resources will become more significant. For this reason, it is quite necessary to implement an in-depth study on changes in spatio-temporal characteristics of glaciers and water resources in recent period and make a comprehensive analysis of the causes. At the same time, apply climate system model to examine the driving factors of glaciers and water resources on different time scales from the perspective of dynamics, predict the future trend in various emission situations, and the relevant achievements can directly serve the major issues of social development such as the formulation of national macro-economic development strategies.

As a developing country that still taking agriculture as a main industry, along with continuous process of industrialization and urbanization as well as the reduction of effective arable land, China's food security will become increasingly prominent; meanwhile, the pressure on ecological and environmental system brought about by rapid economic development will be increasingly serious. Under the circumstance of exacerbated global warming, the impact of climate change on China's agriculture and ecosystem will gradually increase and will eventually affect the sustainability of the healthy development of national economy. Currently, some western developed countries have established assessment models of agricultural, ecological and economic impacts relying on climate system model through comprehensive analysis of the information in the fields such as climate change, farm produce yield and various ecosystem indexes, for assessing the potential influence of climate change on the above-mentioned areas. Based on interdisciplinary cooperation, China is in urgent need of making deployment for those issues mentioned above.

4. Industrial and Energy Consumption Restructuring of Greenhouse Gas Emission Reduction and the Key Technologies

The following aspects require special research: restructuring of industrial and energy consumption, development of resource-conserving economic development model targeting at reducing greenhouse gas emission; technical standards and tax polices of greenhouse gas emission for energy-consuming enterprises; accumulation of basic data that serving national greenhouse gas management and carbon sequestration inventory as well as comprehensive management of terrestrial carbon sequestration; research on agricultural nitrous oxide and paddy field methane emissions, fixed technology system of agricultural soils towards atmospheric carbon dioxide and the integrated application of bio-technologies such as no tillage or minimum tillage, balanced fertilizer application, water-saving irrigation and the management of soil bio-organic fertilizer.

5. The Adaptability of Social Economy and Ecosystem to Climate Change and Its Management

Human activity is a major factor in climate change, but also the outcome of long-term adaptation to climate change. The climate change in the past century produced a certain impact on terrestrial ecosystem and global socio-economic system, and it may generate a greater impact in the next century. Therefore, it is a must for us to study and assess the major issues such as the impact of climate change on ecological systems, freshwater resources, productivity of agriculture and forestry, food security in China and the surrounding regions; the impact of climate change on the pattern, process and service function of terrestrial ecosystem, and on the urbanization layout, city ecosystem structure and function design; the impact of climate change on disease spread and human health; the impact of climate change on the vulnerability and adaptability of socio-economic system as well as management risk, and so on.

Adaptation to climate change is an inevitable choice of human beings; people expect to slow the climate change process, adapt to the outcome of climate change and lower the social, economic and ecological risks in human development through adjusting their behavior, life, production mode and economic growth model. Therefore, it is necessary to study China's energy production and consumption, industrial production process, potential, methods and technologies of reducing greenhouse gas emission in agriculture and forestry production; research China's construction of agricultural infrastructure and scientific planning of grain base in coping with climate change and strengthening the scale of advantageous produce production and regional crop distribution; study the design standards of water infrastructure and moisture-proof facilities of China in response to climate change; research the management of afforestation technology, ecosystem, forestry and grassland adaptive to climate change, as well as the potential, means, methods of bio-energy development and safety assessment system; study the adaptation of social economy and ecosystem towards climate change as well as the system of risk management technology.

6. Climate Change in Typical or Key Regions of China and the Response of Their Ecosystems

China has a vast territory; some special geographical units constitute the key of influence on climate change and those areas are also climate change sensitive regions. The study on the influence of such areas on climate change and the response of their ecosystems to climate change not only has global scientific significance, but is also the scientific and technological need of China to cope with climate change and eco-building. Emphasis should be laid on the research of climate change characteristics of the Qinghai-Tibet Plateau and the impact on global climate change as well as the regional sensitivity; the interaction between climate system and the cryosphere of China, the response of cryosphere ecosystem of China and its adjacent areas to climate change and the potential socio-economic impact; the process and impact of temporal and spatial variation of East China monsoon region; the impact of climate change on ecosystem, biodiversity and ecological functions of the forest in Northeast China; the dry warming process mechanism and trend in North China as well as the impact on water resources, agricultural production and socio-economic development; the driving mechanism of human activities and climate change on desertification process; climate theory of the occurrence, development and regional transmission of dust storm; the impact of climate change on energy exchange, moisture and nutrient cycling of basin-scale mountain - oasis - desert complex ecosystem and its response mechanism; the impact of climate change and sea level rise on coastal urban planning, infrastructure and socio-economic development and the countermeasures.

7. The Response of Ecosystem to Global Climate Change and the Adaptive Environmental and Biological Control Mechanism

The response of ecosystem to global climate change and the adaptive environmental and biological control mechanism are the theoretical basis for recognizing and predicting the impact of climate change on ecosystem and also the scientific basis for formulating ecosystem adaptive management. Hence, there is a need for organizing the response and adaptability network pilot study of ecosystem on global climate change on various scales such as ecosystem (station), transect and region, etc., to uncover the adaptation mechanism of ecosystem to climate change as well as its feedback mechanism to global climate change. The key research points include: ① take the typical forest, grassland and agro-ecological system as the main objects, carry out environmental factor control experiments, such as simulated nitrogen deposition, open-top chamber warming trial, free air carbon dioxide enrichment and artificial precipitation, etc., analyze the impact of key factors relevant to climate change on ecosystem structure and function (such as temperature rise, changes in precipitation patterns, concentration changes of atmospheric carbon dioxide and ozone and nitrogen deposition, etc.), and clarify the response and adaptation mechanism of carbon, nitrogen and water cycle processes in ecosystem, in particular the process of soil, as well as the coupling relationship between ground and underground cycles, to global change; ② take the farming-grazing transitional zone in North China, Northeast China forest-steppe ecozone and the Qinghai-Tibet Plateau, etc., as the main study objects to research the ecosystem structure, function and process changes under the dual roles of climate change and human activities, as well as the mechanism of response and adaptation; ③ research and assess the impact of extreme weather events and natural disasters on terrestrial ecosystem biogeochemical cycles as well as the ecosystem response and adaptation laws by means of site investigation, flux observation, physiological and ecological field experiment, remote sensing monitoring and simulation and modeling, etc., to explore the theory of ecosystem adaptability management.

8. Scientific Principles of Carbon-nitrogen Cycle of Terrestrial Ecosystems and Greenhouse Gas Management

The carbon-nitrogen cycle and water cycle of terrestrial ecosystems are mutually coupled. The three major ecological processes are closely related to climate change. Carrying out network observation of the fluxes of carbon, nitrogen and water of terrestrial ecosystems as well as their cycling processes, organizing relevant and adaptive field control tests of the response of the key processes of carbon-nitrogen cycle to climate change have been the core scientific areas of ecological systems and climate change research. The research priorities of China should include: ① to build a comprehensive observation system on fluxes of carbon, nitrogen and water as well as their coupling cycling processes covering all types of major terrestrial ecological systems (forests, grasslands, farmlands, wetlands and cities) in China, based on the observation and research networks of carbon dioxide, nitrous oxide, methane, water vapor and energy flux of terrestrial ecosystems and terrestrial transects in China. By combining the observation of greenhouse gas flux and field experiments of ecological processes with technologies and methods of stable isotope, numerical simulation and remote sensing, to reveal the cycling of carbon, nitrogen and water in eco-systems and the regional characteristics of their coupling relations as well as the control mechanism of biology and environment and to clarify the metabolic characteristics, efficiency of use and laws of physiological and ecological balances (including laws of distribution, transformation and transfer) of major life elements of ecological systems (carbon, nitrogen, potassium, phosphorus, etc.); ② to study and develop multi-scale data observation-model systems and mechanism models, and to establish national measurement system and decision supporting system for the management of greenhouse gases in terrestrial ecosystems and provide scientific and technological support for China's fulfillment of *United Nations Framework Convention on Climate Change* and the management of greenhouse gases and water resource ; ③ to use ground data and remote sensing observational data of climate, vegetation, soil, carbon flux with ecosystem process and mechanism model as well as remote sensing model to conduct quantitative analysis for carbon source and sequestration dynamics in typical regions or main terrestrial ecosystems, and evaluate the contribution of China's terrestrial ecosystems to the carbon balances of East Asia and the world; ④ to study the ecological effects and carbon sequestration function of China's major construction projects (reforestation of the cultivated land, turning grazing land to grassland, natural forests protection, shelter forests construction); ⑤ to study the absorption and emission, capture and underground sequestration of greenhouse gases of ecosystems, reducing the emissions of greenhouse gases in the production of agriculture and animal husbandry, and principle of technologies of biomass resources and energy application or other terrestrial carbon sequestration adding technologies.

9. International Cooperation in Climate Change and Strategies and Tactics for Diplomatic Negotiations

As a big developing country, China has caught world wide attention for its economic development mode and achievements since it implemented the reforming and opening up policies. At the same time, global climate change has been a core issue in international community in recent years.

China has encountered many environmental problems brought by the rapid economic development in the last two decades and the pressure from international community has kept increasing.

So far, China has successively signed *United Nations Framework*

Convention on Climate Change, *United Nations Convention to Combat Desertification*, *Ramsar Convention*, *Convention on Biodiversity*, *Stockholm Convention on Implementing International Action on Certain Persistent Organic Pollutants* and other international conventions regarding global climate change. With the responsibility of emission reduction and the increasing pressure of economic development layout adjustment, plus the continual emergence of new problems in international community in the background of climate change, China's environmental diplomatic problems will become even more prominent. As a responsible developing country, China currently needs to set up state-level inter-disciplinary research team, through collaborative research to accurately grasp the histories, status quo, and future development trends of economies, ecology and environment of China and other countries. Especially, China has the discourse power on issues related to China's regional climate change in a scientific sense, thus China needs to develop feasible measures in terms of international cooperation on climate change and strategies for diplomatic negotiations and needs to make scientific international cooperation and strategies for diplomatic negotiations by considering China's responsibility and the protection of national interests to the greatest extent as soon as possible.

The roadmap of science and technology development in this field and main scientific issues and key technologies are shown in Fig. 7.1 and Table 7.1.

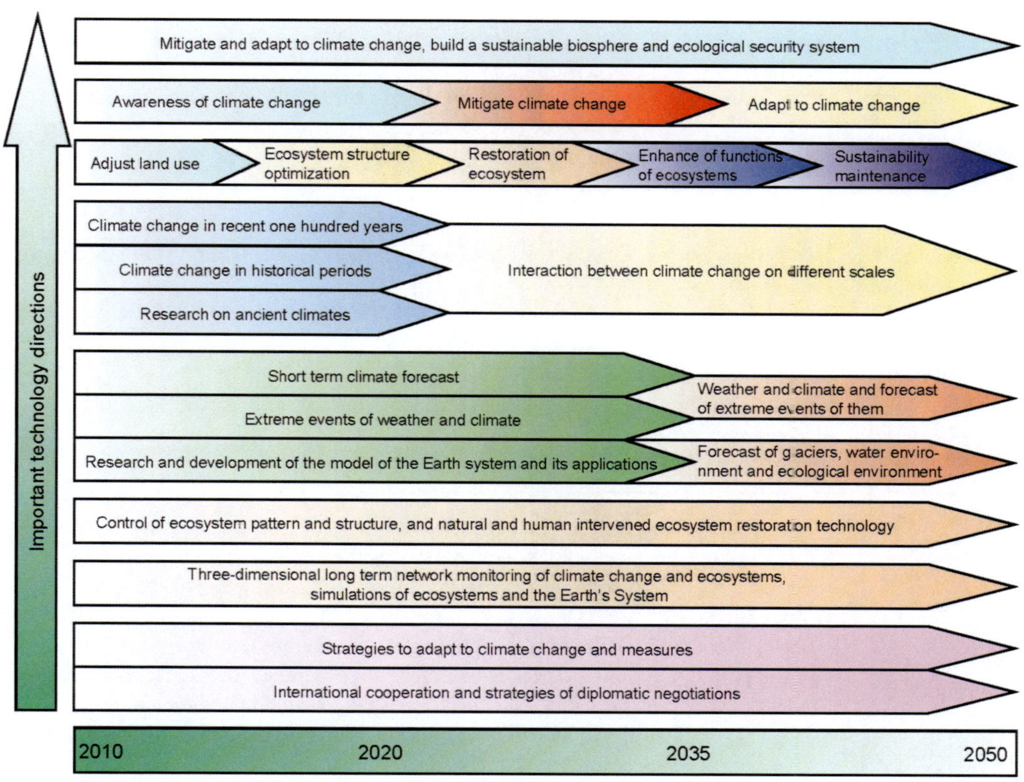

Fig. 7.1 Global climate change and roadmap of science and technology development in ecological processes

Table 7.1 Main scientific issues and key technologies in global climate change and its ecological processes

Main Objectives	Scientific and technological areas	Main scientific issues and key technologies
Reveal the facts of China's climate change and the causes in the background of global warming, develop China's Earth System models and increase forecasting and estimation accuracy for weather, climate and extreme events, and serve for the needs of the country	Environmental science, Climatology	Paleoclimate studies: the rehabilitation and reconstruction of ancient climate, ancient climate modeling for typical periods and phases
	Environmental science, Climatology	Climate changes in historical periods: climate reconstruction in historical periods in China, climate simulations for the past one thousand years, and causes for climates of historical periods of China
	Environmental science, Climatology	The facts and the causes of climate change in China for the past one hundred years: inter-decadal climate change, correlation between seasonal and decadal-scale climate change, extreme weather and climate events research
	Environmental science, Climatology	Earth system model development and applications: multi-disciplinary teams jointly develop the model of China's Earth system, strengthen its application, and study on the Earth's climate change from kinetics mechanism
	Environmental science, Climatology	Short-term weather forecasting, climate forecasts for decades to a century-scale: climate forecasting for seasonal to annual scale, climate forecasting for the next decade to a century-scale, forecasting for extreme weather and climate events, assessment of changes in glaciers and water resources, assessment of impacts of climate change on agriculture, ecosystems, society and economy
Take international responsibility in reducing emissions, at the same time, protect national interests to the greatest extent	Environmental science, Ecology, Geography	Overall responses and measures of adaptation to climate change
	Environmental science, Ecology, Geography	International cooperation in climate change and an diplomatic negotiations strategies
Greenhouse gas emission reduction industry and energy consumption restructuring and key technologies	Environmental science, Ecology, Geography	Industry and energy consumption structure adjustments, technology issue of development of a resource-conserving economic development mode for the purpose of reducing greenhouse gas emissions, technical standards and tax policy issues for greenhouse gas emissions of energy consuming enterprises, development of new energy-saving technologies for industrial and agricultural production and their applications, development of alternative new technologies for fossil energy and their applications

(Continued)

Main Objectives	Scientific and technological areas	Main scientific issues and key technologies
Adaption of society, economy and ecosystems to climate change and its management	Environmental science, Ecology, Geography	Impacts of climate change on freshwater resources, productivities of agriculture, animal husbandry and forestry of our country and the surrounding regions, food security, disease prevention and control Impacts of climate change on the structures of ecosystems and their functions and the driving mechanisms, impacts of climate change on urban ecosystems and adaptation measures, evidence of impacts of climate change on the health of China's population, risk and adaptation measures, vulnerability and adaptation capacity of impacts of climate change on socio-economic system Agricultural infrastructure, scientific planning for food base, strengthening large-scale crop cultivation of superior agricultural products and regional layout of agricultural products responding to climate change Water infrastructure and design standards for moisture-proof facilities responding to climate change The potential, ways and methods of reducing emission of greenhouse gases in China's energy production and consumption, industrial processes, agriculture and forestry production and the evaluation system, afforestation technology, forestry and grassland management that enhance adaptation capacity of ecosystems and forestry production responding to climate change, potential, means and methods of bio-energy development and safety evaluation system
Climate changes in China's typical or key regions and the responses of their ecosystems	Environmental science, Ecology, Geography	Climate change in Qinghai-Tibet Plateau and its sensitivity, interaction between climate and cryosphere and its potential impacts on China and the adjacent regions, temporal and spatial variation of climate change process in monsoon regions of eastern China and its impact Impacts of climate change on forest ecosystem structure, biological diversity and ecological functions of northeastern China, causes, mechanism, development trends of the process of drying and warming of northeastern China's climate and its impacts on water resources, agricultural production and social and economic development Driving mechanism of impacts of human activities and climate change on desertification process, climatology theory of the occurrence, development and regional transmission of dust storms, impacts of climate change on energy exchange, water and nutrient cycling of basin-scale mountain-oasis-desert complex ecosystem and its response mechanism Impacts of climate change and sea level rise on coastal city planning, infrastructure and socio-economic development and adaptive counter measures
Carbon cycle of ecosystems and the scientific principles of their greenhouse gases management	Environmental science, Ecology, Geography	Carbon and nitrogen cycling mechanism of ecosystems and its relationship with climate change, theory and technology of absorption, emission, capture and underground sequestration of greenhouse gases of ecosystems Greenhouse gases reduction mechanism of agriculture and animal husbandry, mechanism and technologies of terrestrial biological carbon sequestration, biological material resources and energy development and utilization

7.2 Restoration of Degraded Ecosystems, Reconstruction of Ecosystems and Conservation of Biodiversity

Restoration of degraded ecosystems and reconstruction of ecosystems is a complex systematic project. Although there are some researches and explorations about theories and methods of restoration and reconstruction of ecosystems, no theoretical system and technology system have been formed. There is still lack of in-depth studies on basic theories. Five issues of the discipline of restoration of ecosystems have been identified by analyzing the restoration and reconstruction of ecosystems and its applications in the past 50 years both at home and abroad.

Firstly, although researchers have understood the overall framework of ecosystem degradation, they are weak in explaining and researching on the causes and factors of ecosystem degradation and its mechanism, which are key research focuses and core research issues needing in-depth exploration. Secondly, currently the exemplary experimental research on restoration and reconstruction of degraded ecosystems is limited to some small, local or regional scope and single community or vegetation types, lacking comprehensive study on overall river basins or on systematic level regional scale, also lacking control strategies optimization researches on existing mode changing with time, global change and the needs of economic and social development. Thirdly, the ultimate goal of restoration and reconstruction of degraded ecosystems is to protect the self-maintaining status after restoration, which requires the establishment of ecological sustainability indicators and methods of assessment. However, there is no effective and accurate quantitative method of diagnosis of degraded ecosystems and no operational indicators and models for monitoring, evaluation of ecosystems and their restoration and reconstruction. Fourthly, the economic behaviors in the process of restoration and reconstruction of ecosystems are essential to successful restoration and reconstruction. However, there is little research on economic behaviors in the process of restoration and reconstruction of ecosystems. How to realize sustainable socio-economic development of river basins in the restoration and reconstruction of natural-economic-social complex ecosystem is an inevitable issue encountered in a large-scale restoration and reconstruction of ecosystems. Fifthly, theoretical system and technological system of restoration and reconstruction of ecosystems have not been formed yet. The lack of basic theory will result in blindness and uncertainties in the applications of technologies and methods of restoration and reconstruction.

Today, the situation of degradation of ecosystems is general and serious, thus, identifying major types and causes of ecosystems degradation, vegetation recovery mechanism and means of different types of degraded ecosystems and the recovery extent, developing a perfect diagnostic indicator system for recovery of degraded ecosystems in order to maintain sustainable ecological economic development, explaining responses of restoration of ecosystems to regional environment and global change, proposing macro-regulation strategies and the overall deployment and means of ecological environment construction, expediting the development of theoretical system and technology system of restoration and reconstruction of ecosystems with Chinese characteristics, and promoting China's natural, social and economic sustainable development is a very tough task. The key scientific and technological issues are as follows.

1. The Causes of Degraded Ecosystem, Restoration Mechanism and Ways to Maintain Stability

Propose complete classification evaluation indicator system and methods, complete a nationwide survey of degraded ecosystems, develop a classification system for degraded ecosystems, provide basic data for researches and treatment on degraded ecosystems and provide basic data for long-term monitoring of ecology and environment. Study on causing mechanism, dynamic process and development trends of all kinds of ecosystem degradation types, propose macro-regulation strategies for degraded ecosystems and overall deployment and means of ecological environment construction as well as research and develop comprehensive diagnostic indicator system for restoration of degraded ecosystems.

Carry out researches on all kinds of ecosystem degradation types and their causes, vegetation recovery mechanism and means of degraded ecosystems in areas with different ecological types and their recovery extents. Researches need to focus on human intervention conditions and methods for natural vegetation recovery processes in China's different types of ecosystems, the potential of vegetation recovery in different types of ecosystems, stability maintaining mechanism, coupled relationships between regional vegetation in different types of ecosystems and environmental factors, ecological function evaluation standards for vegetation in different regions, distribution pattern of vegetation in different types of ecosystems and landscape pattern in different scales and their relationships with ecosystems, etc.

2. Integrated Research on Surface Process of Terrestrial Ecosystems and its Pattern

A terrestrial surface ecosystem includes structures, evolution, development and interactions between environment, resources and social economy which are closely related to human in time and space, which is the largest sub-system impacted by human activities on the surface of the earth. Thus, studying on it is an important area for future restoration ecology development. Researches on terrestrial surface ecological process will develop toward both macro and micro directions. The key area is comprehensive research on various surface processes of ecosystems. Recent major research areas include: hydrological cycle and water flow transformation mechanism and its physical experimental simulations; material transfer process of ecosystems in river basins and its impacts on regional environment; occurrence, development process and driving mechanism of erosion in different time and space scales; land utilization/vegetation evolution process and its regulation mechanism; biogenic elements cycling process and regional-scale conversion in soil - plant -

atmosphere continuous system; environmental biogeochemical processes of life elements, pollutants, heavy metals and rare earths and their health effects.

3. Socio-economic Research on Restoration and Reconstruction of Degraded Ecosystems

Restoration and reconstruction of degraded ecosystems involves a wide range of humanity, social and economic problems. We should carry out researches on restoration and reconstruction of degraded ecosystems in fields of law, environmental science, economics and sociology based on researches on natural sciences. Sometimes, unpredictable practical problems may happen due to the complexity of ecosystems and unexpected environmental factors. Thus, it is needed to conduct research on decision system of restoration and reconstruction of degraded ecosystems and establish the decision framework of restoration and reconstruction planning to conduct decision analysis, risk assessment and management; conduct multi-disciplinary comprehensive and integrated researches on risks, inputs, effects and benefits in restoration process of degraded ecosystems. At the same time, it is needed to strengthen legislative research on restoration of degraded ecosystems, because regional and large scale effective ecological recovery can not rely upon researches of scholars and the passion of the people only. It is required to establish the basis through legislation and guarantee effective implementation through judicial authority.

4. Ecological Research on Dynamics of Populations of Important Biological Resources and the Potentials of their Economic Development

Modern biotechnology industry is an ecological and economic process of promoting biological resources production of biological products required by human to the greatest extent, based on a good ecological environment created artificially. Whether biotechnology industry can be developed to a large-scale and how certain amount of biomass can be established efficiently are the core issues of modern biotechnology industry development. Therefore, we should analyze whether biological resources have potentials for economic development and its required habitat elements for maximizing the output of economic products. Research on the biological characteristics of important biological species and their adaptive potentials, especially, species dynamic succession to obtain optimal sustainable yield and other parameters is the scientific foundation of reasonable and effective exploitation of biological resources, which has an important guiding significance to restoration of regional degraded ecosystems.

5. Impacts of Restoration of Degraded Ecosystems on Regional Environment and Responses to Global Environmental Change

In considering current status of China's degraded ecosystems and the future requirements of restoration and reconstruction, we should strengthen researches on land utilization/vegetation cover change and its driving mechanism, impacts of ecosystem degradation and its restoration on regional ecological environment, responses of development and evolution of degraded ecosystems to global change, adaption of restoration and reconstruction activities of degraded ecosystem to global change, impacts of global environmental change on the structure and functions of ecosystems in vulnerable regions, ecological strategies and countermeasures for global environmental change, etc. In which, social and economic strategies and countermeasures for global environmental change involve natural belt shift change, land utilization and structures and layouts of agriculture, forestry, and animal husbandry, energy restructuring, defense measures of coastal areas, reasonable exploitation of natural resources, natural disasters prevention, etc.

6. Survey of Biodiversity and Monitoring of Alien Invasive Species

Biodiversity resources are strategic resources related to national security for a country to realize sustainable development. Identifying the basic situation of China's biodiversity is of great significance for biodiversity conservation, monitoring alien species invasion and reasonable exploitation of biological resources. Recent research focuses include: in-depth survey of species diversity in China's typical ecosystems and key regions, collection of wildlife resources, developing rapid identification, evaluation and preservation technologies, exploring key characteristics and genes, building digital systems for biodiversity; studying on distribution law of biodiversity, alien species invasion and ecological effects, developing inspection system for species invasion, realizing control of harmful species; fully carrying out researches on China's species resources evaluation, rapid identification, reservation technology and its applications and measures. According to China's great needs of special genetic resources in the future, mainly carrying out mining, conservation and optimal utilization of strategic genetic resources in typical ecosystems and special habitats; rapid evaluation for economic values of biodiversity. At the first stage, we need to take measures for hotspots of China's biodiversity; at the second stage, take measures for national typical regions and neighboring countries.

7. Spatial and Temporal Principles of Biodiversity Evolution and Control of Alien Species Invasion

Research on biodiversity formation and evolution mechanism is the basis of effective protection and scientific utilization of biodiversity resources, which can not only provide guidance of core theories for restoration of degraded ecosystems and reconstruction projects, but also strengthen China's forecast ability on biodiversity and spatial and temporal evolution trends of ecosystems, therefore it can improve the forwardness and scientificalness of critical decisions. Recent major research areas include: distribution patterns of biodiversity in China's different geographical regions and their formation and evolution mechanism; development trends of biodiversity under global change; relationship between biodiversity distribution and historical ecological environmental change; impacts of biodiversity on stability of ecosystems and their functions; origins of critical traits and formation of species and adaptive differentiation; *ex-situ* conservation of endangered species and key technologies of species rejuvenation.

The roadmap of science and technology development in this field and main scientific issues and key technologies are shown in Fig. 7.2 and Table 7.1.

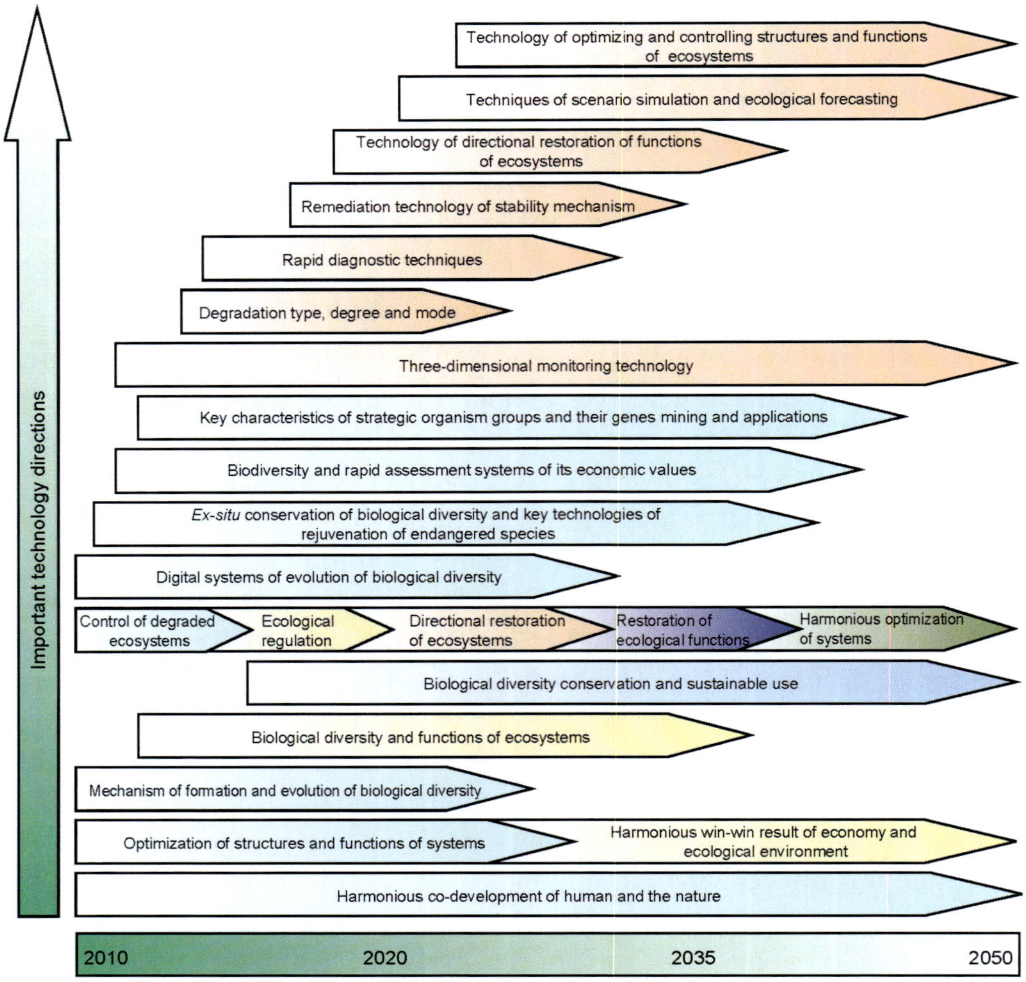

Fig. 7.2 Restoration of degraded ecosystems and roadmap for biodiversity technology development

Table 7.2 Restoration of degraded ecosystems and main scientific issues and key technologies in the area of biodiversity

Main objectives	Scientific and technological areas	Main scientific issues and key technologies
Optimization of structure of ecosystem and its function in harmony	Ecological sciences	Strengthen building of field stations, improve monitoring system of ecological degradation and restoration; causes of degraded ecosystems and their restoration mechanism and means, focus on conditions and methods of human intervention in natural recovery process of different types of vegetation; comprehensive study on surface process of ecosystem and its pattern; research on population dynamics of important biological resources and its potential of economic development; impacts of restoration of degraded ecosystems on regional environments and their responses to global environmental change; key technologies of rapid reconstruction of vegetation and ecological restoration of degraded ecosystems

(Continued)

Main objectives	Scientific and technological areas	Main scientific issues and key technologies
Win-win between social economy and eco-environment	Ecological economics	Socio-economic research of restoration of degraded ecosystems; strengthen legislative research on restoration of degraded ecosystems; develop comprehensive decision supporting system of developing ecological restoration and economy in harmony
Survey of China's biodiversity and its evolution trend	Ecology	Discovering, conservation and reasonable utilization of special genetic resources of China's typical ecosystems; in the first phase, for hotspots of China's biodiversity, in the second phase, scope extended to China's typical regions and neighboring countries
Temporal and spatial principles of evolution of biodiversity and control of alien species invasion	Ecology	Study on evaluation and conservation technologies and utilization strategies of China's biological germplasm resources; collection, evaluation and conservation technologies of wildlife resources and gene discovery; alien species invasion and its ecological effects; relationship between biodiversity and global change; distribution of biodiversity and its migration trends; conservation of biodiversity and regional eco-environments
Comprehensive technology of conservation of biodiversity of vulnerable ecosystems	Ecology	Coupling mechanism of biodiversity and system stability of vulnerable ecosystems, in particular study on underground ecological processes; comprehensive technologies of conservation of vulnerable ecosystem / biodiversity as well as their demonstration and applications

7.3 Urbanization and Environmental Quality

China has entered a period of rapid urbanization. In order to realize harmonious development of urbanization and cities, relevant science and technologies are needed urgently, which include: to develop key technologies and dynamic monitoring technologies of modern urban areas planning, based on understanding future China's urbanization principles, focusing on multidisciplinary scientific planning of urban areas; to clarify complex pollution occurrence mechanism and advanced control technologies; to make breakthrough in directional control technologies of urban waste reclamation and biogeochemical process, develop resource-saving green building materials with high durability, and improve utilization efficiency of urban resources and energy; to strengthen applications of information technologies, develop urban-digital integration management technologies, and build integrated urban management technologies system with high efficiency and multi-functions; to develop urban eco-living environment and construct green buildings; to develop technologies of harmless disposal and reclamation of urban waste water and garbage, and to develop urban residential areas and indoor environment improvement technologies to significantly improve the quality of living environment.

2050

1. Forecasts of the Characteristics of China's Urbanization in Different Stages of its Development from now to 2050

Urbanization is an important symbol of industrialization and modernization, which reflects the progress and level of economic and social development. The key to realize urban and rural co-ordination and solve the problems of "dual structure" and "three rural issues" that have troubled China's economic and social healthy development is to vigorously promote the process of urbanization. Overall, urbanization lagging behind industrialization is still the principal contradiction that restricts rapid economic and social development in China. The next 50 years is an important period for China's urbanization while urbanization is the key driving factor for ecological and environmental evolution. In order to continuously promote the strategy of urbanization, overcome difficult problems emerging in the process of urbanization, prevent other new problems emerging in future process of urbanization, it is necessary for us to explore the characteristics of China's urbanization in different regions and periods and their metabolic characteristics, including social and economic characteristics, social behaviors and urban welfare; to analyze and forecast changes of population structure, industrial structure, production methods and consuming pattern in different stages of urbanization, and provide solutions and measures for implementing a healthy and effective urbanization process.

2. The Process of Urban Ecosystem and Human Stress Mechanism

Cities are areas in which human and ecological environment interact with each other to the strongest degree and also complex ecosystems consisting of human activities and natural factors. The structure and process of urban ecosystems have their own laws. As mankind is lacking of understanding of urban ecosystems and violates ecological laws in the process of urban construction, a series of urban ecological and environmental problems have occurred. Rapid urbanization has resulted in resource depletion, environmental pollution, heat island effect, crowding effect, deterioration of living environment, public health problems and stress to regional ecosystems which constitute critical challenges to China's building a well-off society in future and realization of sustainable development. The sustainability of cities' development determines the sustainability of China's economic and social development to some extent. China is a large regional country. Although the economies of different regions of China are growing with a faster speed than that of before reforming, economies of different regions grow at different speeds and the situation of unbalanced regional economic development has exacerbated. The level of development in eastern regions is much more than that of central and western regions. Analyzing urban ecosystems' patterns and functions of China's eastern and central and western regions in different stages of economic development, systematically exploring the dynamics mechanism of interaction between human activities and urban ecosystems' ecological serving functions, explaining the drive mechanism of the process of urbanization to the evolution of ecosystems and environments, developing comprehensive analyzing and monitoring methods and technologies for ecosystems, exploring comprehensive ecological integration methods and providing scientific basis for planning and management of cities and regions development are the scientific issues needing to be solved immediately in China's current urban development processes.

3. Combined Pollution Occurring Mechanism in the Process of Urbanization and Ecological and Health Effects

In the process of China's rapid urbanization development in the past 30 years, some important economic regions, have formed complex situations of co-existence of pollutions from point source, line source and diffuse source, mix of municipal and industrial wastes, overlap of various new and old pollutants, interaction of water, air, and soil pollutions. Quality of environment is the judgment of human beings on status of environment and behaviors in accordance with the fundamental interest of human beings and is also one of the basic conditions for human existence, living and development like other resources. Since the industrialization, the quality of environment in which human beings live has changed dramatically and profoundly in century-scale. The root causes of the environmental quality change are emissions of pollutants from human's economic activities and changes of natural environment, i.e. results of impacts of human beings on the nature after the Earth's surface ecosystems entering into humanity period. Therefore, the quality of environment is the basis of coordination of human and environment and is the scientific base for sustainable development. Environmental transport and occurring mechanism of complex pollutants and the evolution of quality of environment in the process of urbanization as well as its ecological and health effects are the core issues of current and future urban environmental sciences. The objectives of research are to analyze multi-media transfer and transformation mechanism and transmission means of pollutants in the process of urbanization, explain the ecological and health effects of pollutants of urban ecosystems, research the coupling mechanism of different pollutants in urban ecosystems and environmental compound pollution formation mechanism, explore means of compound pollution and ecological health effects, as well as impacts of changes of quality of urban environment on quality of residential environment and life & health and its control principle. In the first stage, researches are focused on four coastal urban agglomerations in eastern China (Pearl River Delta urban agglomeration, Yangtze River Delta urban agglomeration, western coast of strait urban agglomeration, urban agglomeration in Bohai Rim area); in the second phase, researches are focused on central urban agglomeration (Changsha, Zhuzhou and Xiangtan City Group, Wuhan urban agglomeration), and gradually covering western regions.

4. Cities' Metabolic Mechanisms and their Regulation Technologies

As an area in which resources are consumed and waste are released intensively, a city's substance and energy metabolic process plays a decisive role in improving its environment. The biogeochemical cycling processes of cities profoundly affect urban, regional and global environmental change, while existed matter circulation model of natural ecosystems is not applicable to urban environmental systems. The ecological essence of urban environmental problems is that substance and energy metabolism stays on temporal and spatial scales, which leads to environmental pollution and accumulation of harmful substances. Research on cities' substance and energy metabolism pattern and the metabolic means and metabolic efficiency and understanding the mechanism of substance accumulation and its depletion on a scale of ecosystems are the prerequisite and basis of diagnosis of cities' illness. From a perspective of cities' metabolism, research on the total volume, efficiency, features, structures, evolution law and trends of cities' substance and energy metabolism is a new research area of urban development both at home and abroad, which will be helpful for solving cities' ecological and environmental problems.

Through studying on characteristics of impacts of different social and economic levels, policies and behavior characteristics on cities' metabolism, we could explain impacts of social and economic level, policies and behavior pattern in the process of urbanization on cities' substance and energy metabolism process, build optimized control pattern of cities' metabolism, develop efficiency evaluation methods for cities' substance and energy metabolism caused by socio-economic level, policies and behavior pattern change, simulate the process of substance and energy metabolism efficiency of the process of urbanization, understand the mechanism of different socio-economic levels, policies and behavior pattern on cities' substance and energy metabolism efficiency as well as explore effective directional control means.

5. Ecological Planning and Design for Cities and Urban Agglomerations Based on the Integration of Multiple Disciplines

Ecological planning of cities and urban agglomerations is the basis of ecological construction of cities and urban agglomerations. For the national conditions of China' s current rapid development of urbanization, we need to conduct researches on ecological planning of boom towns and urban agglomerations, realize reasonable land utilization through ecological planning of cities and urban agglomerations, guide rational distribution of cities and urban agglomerations construction, provide technology system services and demonstration for the construction of infrastructure, selection and layout of industries and consumer behavior of local cities and urban agglomerations, and build environmental-friendly new cities and urban agglomerations systems with harmonious development. Using environmental science, urbanology, sensor telemetry and wireless data transmission technology, developing and applying information engineering, systems engineering, RS, GPS, GIS and other engineering technology, integrating sociology, psychology, history, economics, management science and other social sciences, we could analyze the service functions and characteristics of urban complex ecosystems, study the interrelationship and mechanism among the law of urban development, environmental bearing capacity and ecological services, develop planning theory and methods of modern cities and urban agglomerations integrated socio-economic factors and environmental functions according to requirements of urban environment in different socio-economic development stages.

Faced with this complex giant urban system, the contents of research on urban environment have been deepened from various perspectives of different disciplines on different levels to cross-department, interdisciplinary, integration, and the object of study are changed from urban natural ecosystems to natural-socio-economic complex ecosystems. Please refer to figure 7.3 for the basic framework of ecological planning and design of cities and urban agglomerations.

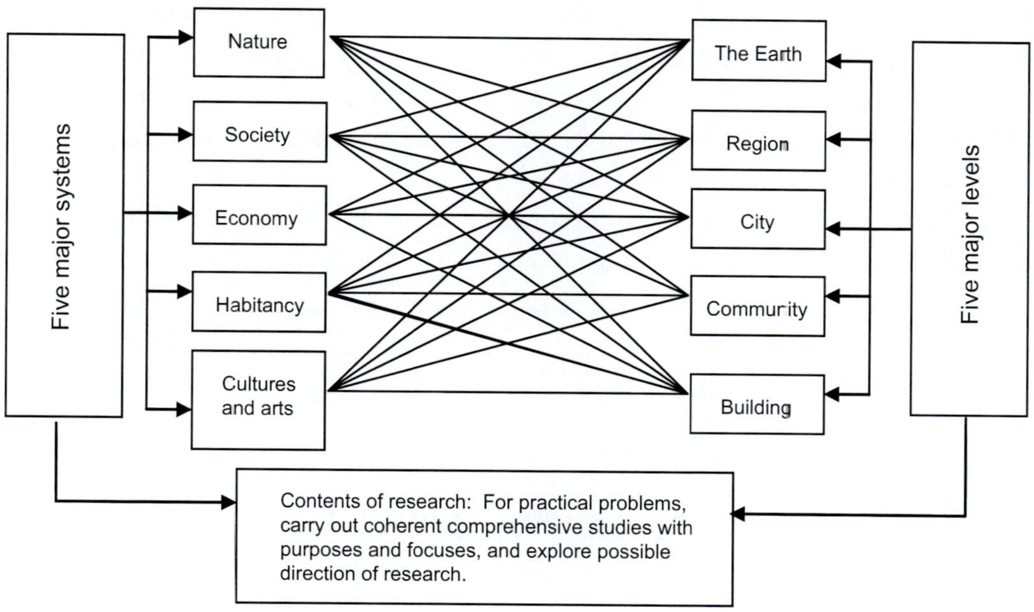

Fig. 7.3 The basic framework of ecological planning and design of cities and urban agglomerations

6. Research on Construction Pattern of Sustainable Cities and Urban Agglomerations

The spread of urban disorder and non-rational growth will make us and our descendants pay high economic, social and environmental costs in future, and the consequence is hard to imagine. In the process of rapid urbanization in China, how to promote the sustainable urban construction comprehensively and effectively is a big problem needing to be solved urgently; the relevant theories and research methods need to be fully concerned and paid more attention to. Through the analysis of the impacts of urban environmental management and policies on urban development and its mechanism, we need to establish urban environmental polices and urban development response models and urban environmental management decision supporting platform, study the environment management mode for sustainable cities, establish theories and methods that could balance sustainable development and environmental protection of different cities and urban agglomerations, explain sustainable production and consuming modes and provide scientific and technological support for China's sustainable construction of cities and urban agglomerations.

Urban agglomeration is an organizational form of regional economic activities. The formation of urban agglomerations and their development is the high-level profile of urbanization. With economic development and the improvement of level of urbanization, urban agglomeration formation and expansion is the trend of urbanization development, which has played an important role in the process of urbanization. In regions of China with higher level of economic development and urbanization, the development and improvement of urban agglomeration is one of the main means of promoting urbanization. Therefore, we need to explore the coupling development and inter-regional coordinate development mechanisms of urban agglomeration, clarify the optimization method of space distribution between urban agglomerations, identify the characteristics of large, middle and small cities of urban agglomerations and their correlations, study regional ecological integration mechanism of urban agglomerations and provide scientific basis for the healthy development of urban agglomerations.

The roadmap of science and technology development in this field and main scientific issues and key technologies are shown in Fig. 7.4 and Table 7.3.

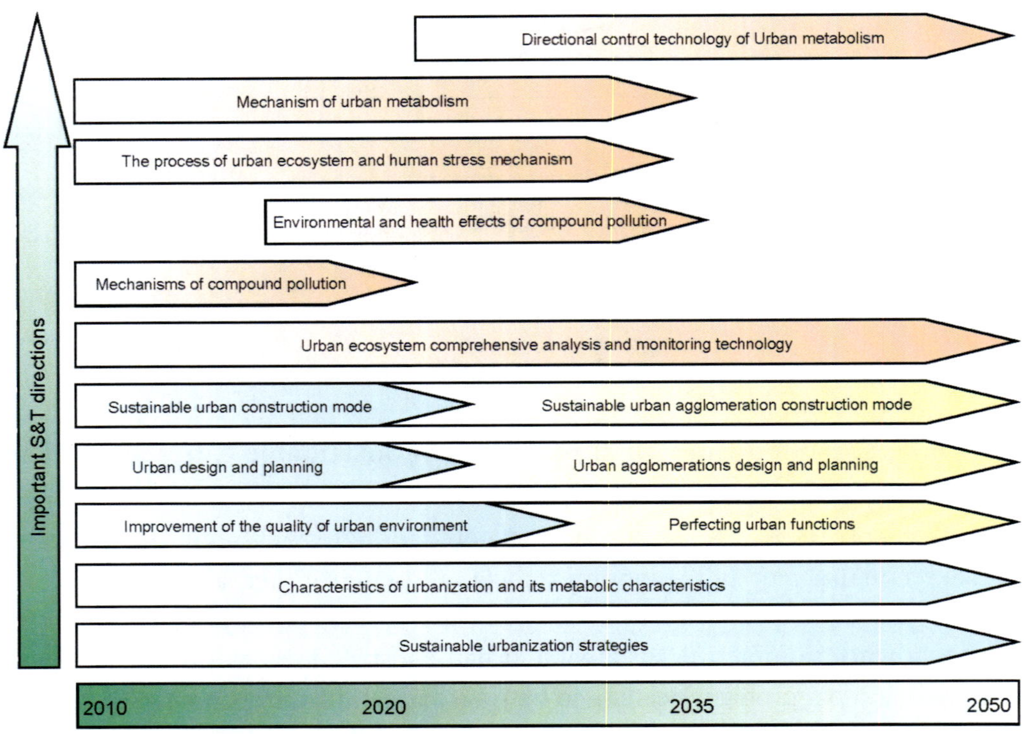

Fig. 7.4 China's S&T roadmap for urbanization and environmental quality to 2050

Table 7.3 Main scientific problems and key technologies in the field of urbanization and environmental quality

Main Objectives	Scientific and technological areas	Main scientific issues and key technologies
Forecast of urbanization and consuming characteristics	Urbanology, environmental science	According to the regional differences of China's urbanization, study urbanization characteristics and consuming characteristics of China's different regions in different periods; analyze and forecast changes of population structure, industrial structure, space expansion, production methods and consuming pattern in different urbanization development stages

(Continued)

Main Objectives	Scientific and technological areas	Main scientific issues and key technologies
The process of urban ecosystems and human stress mechanism	Ecology, environmental science	Study structures and functions of urban ecosystems of eastern, central and western regions of China in different economic development stages, systemically explore dynamics mechanism of interaction between human activities and urban ecosystems service functions, establish comprehensive analysis and monitoring methods and technologies of urban ecosystems
Combined pollution occurring mechanism of urbanization process and its ecological and health effects	Ecology, environmental science	Analyze environmental pollutants multi-media transfer and transformation mechanism and transmission means in the process of urbanization, explain ecological and health effects of pollutants in urban ecosystems; study coupling mechanism of different pollutants in urban ecosystems and formation mechanism of environmental compound pollution and explore formation means of compound pollution and its ecological health effects
Urban metabolism mechanism and regulation technologies	Ecology, environmental science	Clarify impacts of urbanization process on urban substance and energy metabolism process and efficiency of the metabolism assessment methods; simulate substance and energy metabolism efficiency dynamic process; explore effective directional regulation means
Design and planning of cities and urban agglomerations based on integration of multi-disciplines	Urbanology, geography, environmental science and ecology	Analyze service functions and characteristics of compound urban ecosystems, study the interrelationship and mechanism among the law of urban development, environmental bearing capacity and ecological services, develop theories and methods of modern urban planning, coupling socio-economic factors and environmental functions
Research on sustainable cities and urban agglomerations construction mode	Environmental economy and environment management	Analyze impacts of urban environmental management and policies on development of cities and its mechanism, establish urban environmental policies and urban development response model and urban environment management and decision supporting platform, explain sustainable production and consuming pattern; coupling development mechanism of urban agglomerations and inter-regional coordinate development mechanism, explain optimizing methods of space distribution between urban agglomerations and explore integration mechanism of regional ecosystems of urban agglomerations

7.4 Land/River Basins/Coastal Zones Biogeochemistry Processes

In the 21st century, human survival and development are facing many problems. Land as the basic element and carrier of human survival and development, a series of environmental and ecological impacts caused by rapid land use/land cover change has become an important concern of ecological and environmental protection. China's economic growth is heavily dependent on high intensity of land development. Rapid development of industrialization, urbanization and transport infrastructure has led to loss of a large amount of arable land; unreasonable land reclamation and use has resulted in exacerbated soil and water pollution, grassland and forest degradation , loss of biodiversity, aggravation of soil erosion, landslide and debris flow and other disasters. River basins are not only current demographic, economic and urban central areas, but also areas with many problems of ecological degradation and deteriorated environment caused by land use/land cover change. These problems have been extended to coastal areas which are most valuable and vulnerable areas in global ecosystem, with rivers as links. This situation has made coastal areas play a pivotal role in the biogeochemical cycling of carbon, nitrogen, sulfur, phosphorus, trace elements and other elements. This kind of researches focus on land/river basins/ coastal biogeochemical processes, from perspectives of harmonious relationship between human and nature, involving different scales for river basin–coastal systems, revealing process, mechanism and trends of land use/land cover change caused by natural and human factors, explaining evolution laws of biogeochemical cycling of carbon, nitrogen, sulfur, phosphorus, trace elements and other elements caused by it and its possible impacts on succession of ecosystems of river basins and coastal belts, proposing strategies and ways of optimizing land use structure, strengthening production and ecological functions of land, maintaining the health of ecosystems of river basins and coastal belts.

1. Accurate Monitoring of Changes of Land System and Ecological Simulation Assessment

Land not only has the function of food production, but also has the function of maintaining regional ecological balance. Maintaining land's ecological function will be the important objective of future land use and optimization. Accessing to accurate real-time information of changes in land systems and assessing ecological effects of changes in land systems are the basis and premise of optimizing land use. The focus in immediate term is to establish and improve space-based and ground-based (satellite, machines, and ground integrated) technologies and methods of land monitoring system; develop key technologies of resource and environmental monitoring in time and space precision, standard-setting, multi-source data fusion and integration, accurate inversion of key parameters, etc.; accurately understand pattern and process of land system structure, production and ecological functions change in real time; quantitatively reveal dominant influencing factors and dynamics mechanism of changes; study laws of output and migration and transformation of carbon, nitrogen, sulfur, phosphorus and other elements caused by different land use patterns and intensity; develop land system structure, production and ecological function changes forecast model; implement ecological assessment of land use/land cover change and ecological impacts scenario analysis of different land intensive utilization and optimal allocation program. The focus in long term is to investigate soil qualities of different regions (basic elements of environmental chemistry background value) in formation, laws of spatial and temporal distribution and changes; solve problem of identification "threshold" value of land ecological security under the condition of multiple stresses and problem of system health indicators; as well as establish theories and methods systems of accurate land utilization planning and design of ecological security of river basins on different scales.

2. Biogeochemical Cycle of River Basins and its Stress on Ecosystems and its Restoration

Biogeochemical cycle of elements is the key to deeply understand all types of ecosystems degradation in river basins. At present, the prevalence of problems of eutrophication in water of rivers and lakes in river basins and ecosystems degradation is related to excessive emissions of nitrogen, phosphor and other biogenic elements caused by land use/land cover change. The focus in immediate term is to study existing forms and laws of trans-media migration and transfer of nitrogen, phosphor and other nutrient elements of river basins in different media of water, sediment and aquatic organism, as well as impacts of temperature, light, transparency and wave, hydrodynamic flow and other water dynamic elements; explore impacts of nutrient elements on river and lake ecosystems structure succession and function change as well as feedback effects of ecosystems' structures and function changes to biogeochemical process of nutrient elements; study theories and methods of river basins' nutrient elements control and block and restoration of damaged ecosystems. The focus in long term is to study the biogeochemical processes, dynamic mechanism and ecosystem effect of carbon, sulfur and other trace elements released from land use/land cover change, explore healthy "threshold" value of contents of elements in different ecosystems, study and maintain all types of ecosystems' steady-state conversion and integrity as well as theories and technologies of collection and storage of excessive carbon, sulfur and trace elements in different media.

3. Coastal Areas' Biogeochemical Processes and Control

Carbon, nitrogen, sulfur, phosphor and other elements released by land use/land cover change will be finally conveyed to coastal ecosystems through a series of rivers and lakes' migration and transformation processes, which will impact succession of coastal and marine ecosystems. The focus in immediate term is to study changes in biogeochemical processes and coupling of coastal ecosystems and terrestrial, marine and other ecosystems around; quantitative describe laws of internal and external migration and transformation of biogenic material and a variety of elemental speciation in the coastal zone and its dynamics mechanism; taking river basins – coastal areas – open ocean as a whole, by means of integration of different disciplines and research means, comprehensively survey process of biogeochemical cycles of nutrient salt and pollutants in the system and its impact on coastal ecosystems and human's health. The focus in long term is to use a large number of long-term and systematic field observations, combining with physical oceanography, marine ecology, marine biology and other disciplines, and coupling biogeochemical cycles model and primary productivity model to study interaction of biogeochemical cycles of nutrient salt and pollutants and food net in coastal ocean areas under double influences of global climate change and human activities, and its impacts on diversity, ecological function diversity and primary productivity of coastal areas and ocean ecosystems, and propose control strategies for healthy conservation of ecosystems.

4. Interaction Between Land and Ocean and River Basins/Coastal Zones Comprehensive Management Theories and Methods

Through implementation of comprehensive management of river basins/coastal zones, relieving resources and environmental problems has been the consensus of governments, scientific community and general public. The focus is to study land – ocean, ocean–atmosphere, coastal zones – ocean's substance migration and flux exchange between interfaces and its spatial and temporal changes; explore impacts of high intensive human activities and global warming and other factors on the land-ocean interaction; study development limit and safety threshold of water, land, biological, mineral and other resources in river basins/coastal zones; establish zoning principles and methods of river basins/coastal zone's ecological-economic functions; identify different sub-districts' dominant ecological-economic function and scope of important ecological function conservation areas; develop "digital river basins/coastal zones" and virtual reality technologies of river basins/coastal zones; establish quantitative diagnosis and pre-warning assessment model of river basins/coastal health and its comprehensive decisive supporting system of river basins/coastal zones; as well as develop a set of theories and technologies of virtuous biogeochemical circles and ecosystems' health maintenance in river basins/coastal zone.

The roadmap of science and technology development in this field, main scientific issues and key technologies are shown in Fig. 7.5 and Table 7.4.

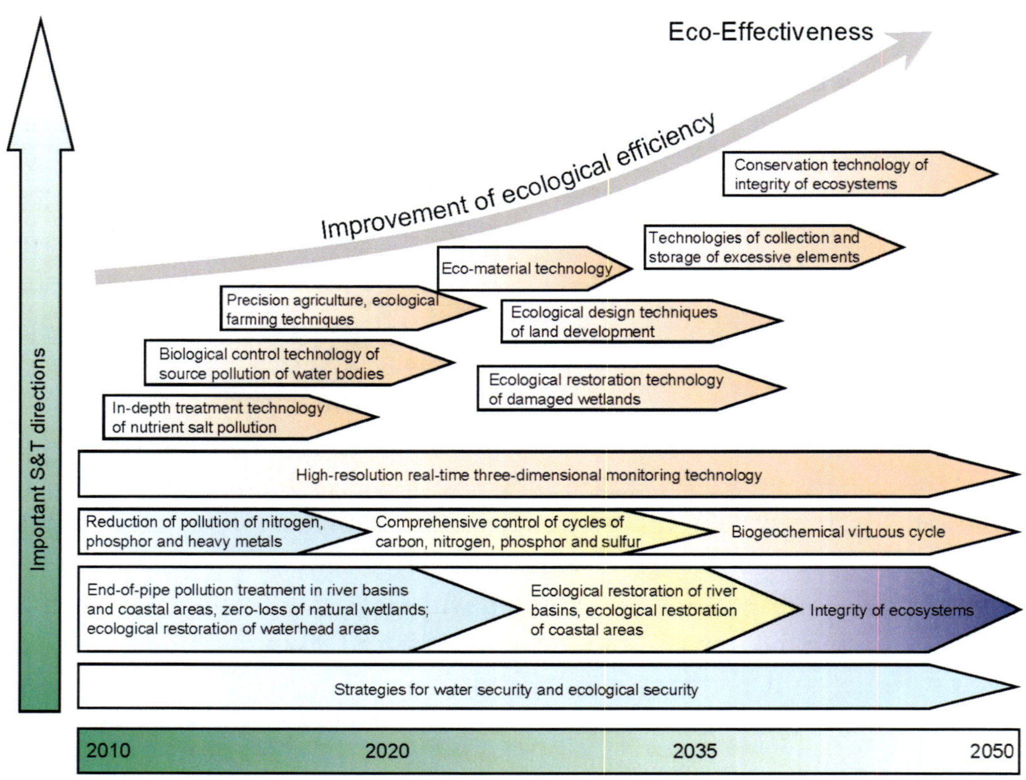

Fig. 7.5 China's S&T roadmap for biogeochemical processes in land / river basins / coastal zones

Table 7.4 Main scientific issues and key technologies of land / river basins / coastal zones biogeochemical processes

Main Objectives	Scientific and technological areas	Main scientific issues and key technologies
Principles of biogeochemical cycles	Geochemistry	Relationship between changes in biogeochemical cycles of carbon, nitrogen, sulfur, phosphorus and trace elements and evolution of land systems, ecological environment; biogeochemical behaviors of key biogenic elements of land systems, river basins/coastal zones interacting with marine systems and their dynamics mechanism; changes in biogeochemical flux of different subsystem and their eco-environmental effects and etc.
Restoration and reconstruction of ecosystem and its integrity protection	Ecological Engineering	Structures, functions and changing processes and patterns of land/ river basins/coastal zone systems and the relationship between them and ecosystem service function; control of all kinds of elements and principle of technologies and methods of blocking their cycle processes; restoration and reconstruction of damaged ecosystems of land, water bodies and coastal zones and means and technical methods of maintaining integrity of system
Drainage area land, ecological environment planning and management	Planning and management	Precise use planning and design theories and methods of land ecological security; zoning principles and methods of river basins/ coastal zone ecology-economic functions; integrity of river basins/ coastal zone ecosystems and optimization methods; mechanism and system of comprehensive management of river basins
Ecological and environmental three-dimensional monitoring and early warning	Information technology	Space-based and ground-based combined integration of land, ecological environment three-dimensional high-precision monitoring techniques; key parameters accurate inversion technologies; changes in systems of land/river basins/coastal zones and their ecological effects simulation and early warning; digital river basins/coastal zones and integrated management decision-making support and etc.

7.5 Control and Restoration of Environmental Pollution

In the past 30 years, China achieved economic development goals to some extent, which the West spent more than 100 years to achieve. Such rapid economic development has brought unprecedented pressure on resource exploitation as well as ecological and environmental protection. Problems are as below: low utilization efficiency of energy and resources, high intensity of pollutants emission, which has exceeded their environmental bearing capacities in many areas; pollution and its stress on ecosystems have spread from land to offshore areas, from surface to underground areas and from single pollution to compound pollution; industrial structural pollution has shown phenomenon of gradient migration and changes on different spatial scales; in some important economic areas and river basins, there is coexistence of pollutions from point source, line source and diffuse source, mix of municipal and industrial wastes, intertwining of all kinds of old and new pollutants, and complex situation of interaction between water, air and soil pollution, as well as potential problems of nuclear and radiation environmental security; a variety of damaged environmental systems need to be restored. Researches on environmental pollution control and restoration should focus on solving mechanism problems of pollution processes and their effects, control and treatment technology problem as well as restoration and repair method problem.

1. Mechanism and Control of Atmospheric Compound Pollution in Urban Agglomerations

Atmospheric compound pollution in urban agglomerations will be a critical environmental problem with direct impacts on human health and social development induced by rapid economic growth in the next 40 to 50 years in China. Understanding and controlling the formation of atmospheric fine particles and oxidants and their compound pollution processes is the core task. Therefore, in the first phase, we should develop monitoring and characterization technologies of atmospheric fine particles and oxidants, at the same time, explore dominant control factors of fine particles and oxidants formation and their interaction mechanisms; in the second phase, for change characteristics of regional atmospheric environments, develop critical and integrated systematic control technologies of fine particles and oxidants, and establish prediction and forecast system of atmospheric environmental quality.

2. Water Pollution Process of River Basins and Security of Water Quality

Shortage of water resources and river basins' water pollution will be a big long-term challenge to China's socio-economic development. To safeguard water quality of all chains of water cycle system will be the fundamental way to solve China's water shortage. In the first phase, for solving compound pollution problems of conventional and toxic and harmful matters in water, we should establish basic database of river basins water quality, explore critical river basins compound pollution processes, develop new theories and new technologies of pollution control as well as develop and integrate drinking water security guarantee and pollution water recycling new technologies at different levels of river basins, critical water sources and water resource comprehensive utilization processes; in the second phase, based on human/ecosystem health risks from water compound pollution, we will focus on river basins' water composite regulation mechanism and build water quality security scientific and technological systems in the process of water cycling.

3. Soil Pollution Control and Technologies of Restoration of Contaminated Sites

In basic research aspect, the main task is to focus on pollutants migration and transformation mechanisms of water – soil – biological systems, verify existence of pollutants in soil and their control factors at molecular level, explain process mechanism and the dominant control factors of pollutants transmitted to water bodies and food chains. In applied research aspect, for China's mineral resources development concentrated areas, we will comprehensively apply ecological design theories to restore and reconstruct damaged ecosystems; for medium and low-polluted farmland, we will develop pollution control chemistry and bio-technologies to realize security of headwaters in food safety. In application aspect, the focus is to develop polluted sites restoration and integration technology, and demonstrate their application in China's typical cities and their suburbs.

4. Multi-media Cycle of Pollutants and Comprehensive Control of Environmental Risks

Pollutants in environment are produced cycled, and transferred in water, air, soil, organism and other multi-media. Therefore, different media have mutual-influencing and interaction characteristics. In the next decades, this kind of complex interaction and influence will be an important research area of environmental science and technological research. To this end, in the first phase, we will select critical river basins or regions to study laws of migration, distribution and transformation of conventional organic matter such as nitrogen, phosphor and representative harmful toxic substances in water, air and soil; propose mechanism of interaction between environmental media taking pollutants as the main clue; explain the joint effect of pollutants under physical, chemical and biological joint process; develop critical control technology. In the second phase, we will establish cycling mode of typical pollutants in different environmental media, and build environmental risk control methods and management system based on multi-media matter circulation and integrity of ecosystems.

The roadmap of science and technology development in this field and main scientific issues and key technologies are shown in Fig. 7.6 and Table 7.5.

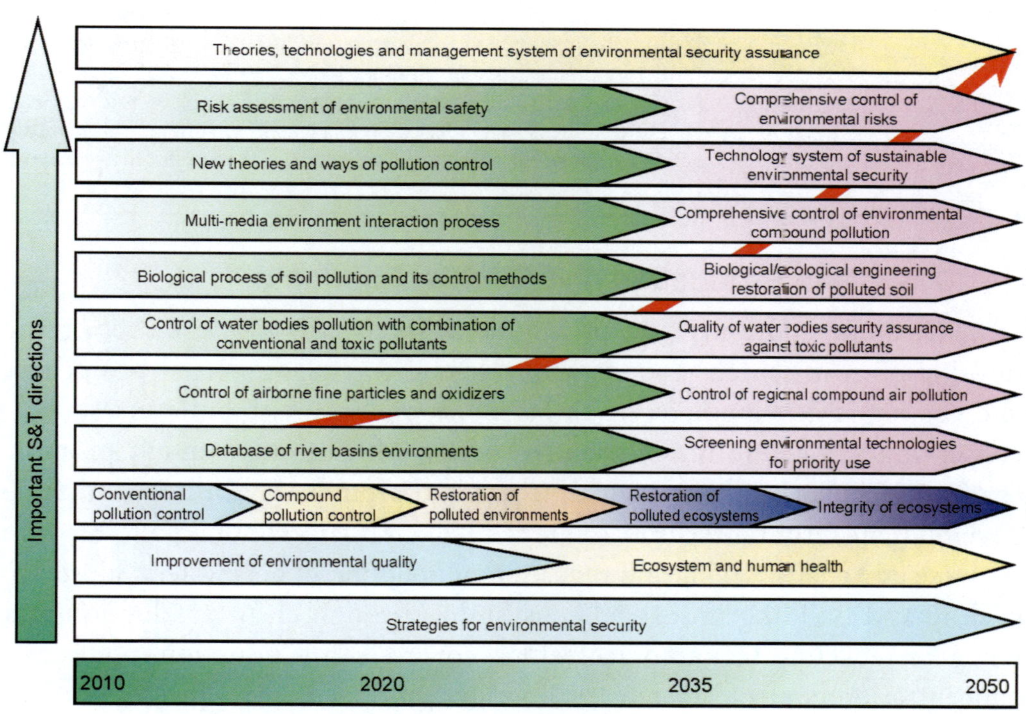

Fig. 7.6 China's S&T roadmap for environmental pollution control and restoration

Table 7.5 Main scientific issues and key technologies of environmental pollution control and restoration

Main Objectives	Scientific and technological areas	Main scientific issues and key technologies
Mechanisms and control of atmospheric pollution in urban agglomerations	Environmental science and engineering, information science	Formation of fine particles and oxidants in atmosphere and the process of their compound pollution; reasons for dust haze weather formation; key driving factors of changes in environmental quality of the air; critical and integration technologies of atmospheric compound pollution control; forecasting technology of atmospheric environmental quality
Pollution process and water quality security of river basins	Environmental science and engineering, ecology, information science	early warning and of monitoring of water quality and basic database of river basins; water environment and benchmark of water quality; process of compound pollution of water bodies, ecological and health effects and principle of control and restoration, drinking water security, and theories and technology system of reuse of wastewater; water cycle security mode based on human/ecosystem health; risk management of water quality in river basins water bodies
Soil pollution control and restoration of polluted sites	Environmental science and engineering, ecology, information science	Rich deposit of pollutants in soil on molecular level and the control factors; mechanism of process of pollutants transmitted to water bodies and the food chains and its major control factors; restoration and reconstruction of damaged ecosystems and its ecological engineering technology; chemical and biological methods of control of pollution in farmland; restoration and demonstration of typical urban and peri-urban areas polluted sites
Multi-media cycle of pollutants and integrated control of environmental risks	Environmental science and engineering, ecology	Multi-media and multi-interface environmental behaviors of typical pollutants; cycles and transformation of pollutants in water, air, soil, organisms; key and integrated technologies of improving environmental quality based on multi-media cycles of substances; integrity of ecosystems based on changes in environmental quality; comprehensive control of environmental quality and risk management

7.6 Clean Production and Circular Economy

Circular economy is a means to achieve new mode of economic development in which resource conservation, environmental protection and economic development are in harmony, and a way to build a resource-conserving and environment-friendly society, which fundamentally promote change in China's economic growth mode, and promote to achieve sustainable development. From a technical perspective, circular economy is to achieve resource and energy saving and reduction of pollution of headwaters, substance cycling and building industrial ecosystem, through clean production and resource recycling technologies, which cover clean production, pollutants concentration control, waste recycling, design of ecological products, eco-industry links, circular economy decision support and other technology systems, whose ultimate goal is to high harmonious coexistence of production process, consuming process and natural ecosystems.

From perspective of development trends of circular economy both at home and abroad, the United States, Japan, EU and other developed countries and regions developed clean production and circular economy after they completed industrial and modernization process, and entered into phase of high-value manufacturing and circular social construction. It mainly focused on resource-saving and pollution control of high-value manufacturing and recycling of social waste, and implemented a series of environment protection-oriented clean production and circular economy critical S&T plan which stimulated a new round of green technology revolution. China is still in the medium term of industrialization, with low efficiency of resource use and severe environmental pollution. Thus, the problems of resources and environment have become the bottleneck of China's economic and social sustainable development. In China's current stage, the main task of development of circular economy should not only include waste recycling, but also focus on upgrading of industrial technologies of source reduction. Based on strengthening end-of-pipe pollution treatment, we should explore ecological transformation of critical chains such as production, circulation and consuming. On 1 January, 2009, China began to implement the *Circular Economy Promotion Law*, which requires governments at all levels to integrate circular economy into their planning, set up a special financial funds (including special science and technology funds), take "reduction, reuse and recycling" as the main means, combine resource conservation, environmental protection with enhancement of economic benefits, realize fundamental change of idea of environmental pollution control, promote upgrading of industrial technologies, and fundamentally safeguard environmental security and sustainable development.

It is predicted that China will basically realize its modernization in 2040 or so. In order to achieve this goal, it is urgently needed to develop circular economy on a large scale and change economic growth pattern. However, generally speaking, China's cleaner production and scientific and technological innovation in area of circular economy are still at the developing stage, has yet to start critical scientific and technological action plan for clean production and circular economy critical similar to that of the United States, Japan, the European Union and other developed countries and regions. China's clean production and circular economy innovation ability can not meet the need of large-scale development of circular economy. It is urgently needed to strengthen scientific and technological innovation top-level design of China's circular economy and its overall layout, establish clean production – integrated control of pollutants – waste recycling – eco-industry links – eco-product design – circular economy decision-making support technologies system, and lead the evolution of clean production and circular economy technological revolution in China.

Clean production and circular economy

1. Clean production and circular economy

The term of clean production was initiated by United Nations Environment

Program, which is a comprehensive and preventative strategy continuously implemented in production process, products and services, in order to increase eco-efficiency and reduce the risks of interaction between human and environments. For production process, it means to save raw materials and energy, reduce the use of toxic materials and reduce its toxicity and amount before a variety of waste emitted from production process; for products, it means to reduce their impacts on human health and environments, in their whole lifecycles–from exploitation of raw materials to final treatment and disposal of discarded products; for service, it means to integrate consideration of impacts on environments in its design and service activities provided.

Circular economy is the deepening of clean production and its development, which can be deemed as an abbreviation of economic growth and development mode based on matter's metabolism process. Up to now, circular economy has been developed into a new economic development mode to achieve resource conservation, environmental protection and economic development in harmony. In accordance with the definition of *Circular Economy Promotion Law* implemented on 1 January, 2009, circular economy refers to the combination of reduction, reuse and recycling activities in the processes of production, circulation and consuming, in which, reduction refers to reduction of resource consumption and recycling waste in the processes of production, circulation and consuming; reuse refers to directly taking waste as products or repair, renovation and reproduce of waste and then continue to use it as product, or use the whole part of waste or partially use waste; recycling refers to use waste as raw material or regenerating waste.

2. Events of China's clean production and circular economy development

In June 2002, Standing Committee of National People's Congress passed *Cleaner Production Promotion Law of the People's Republic of China*, which has laid a basis for China's circular economy legislation.

In July 2005, the State Council issued *Notice of the State Council about recent key work of building saving-oriented society* and *Several suggestion of State Council about speeding up development of circulation economy* (the National Development [2005] No. 21, 22), which systematically explained the guiding ideology of development of circular economy and its basic principles and main objectives, as well as the focus of the development of circular economy and key links.

In October 2005, the National Development and Reform Commission and the State Environmental Protection Administration, Ministry of Science and Technology and other ministries cooperated to perform the first batch of national circular economy pilot projects in key industries and areas, industrial parks and provincial and municipal organizations.

In October 2007, Hu Jintao, General Secretary of the Central Committee of the Communist Party of China, proposed the concept of ecological civilization, and to develop circular economy on a large-scale, form energy resources saving and ecological environment protection industrial structure, growth mode and consuming pattern, in the Report of 17th Representative Conference of the CPC.

In December 2007, the National Development and Reform Commission and the State Environmental Protection Administration, Ministry of Science and Technology and other ministries launched the second batch of pilot projects of circular economy of China. Till now, there have been over 178 experimental units.

In 1 January, 2009, *Circular Economy Promotion Law of the People's Republic of China* was officially implemented, which identified the starting point of the development of circular economy of China, the main direction and implementation ways, and provide a solid legal guarantee.

Source: Zhang Tian-Zhu, Shi Lei, Jia Xiaoping. 2006. Introduction to cleaner production. Beijing: Higher Education Press

In accordance with the national conditions of China at current stage and the medium and long-term development goals, clean production and technologies of circular economy and its practice need to make breakthroughs from the following aspects.

1. Upgrade of Clean Production Technology for High Consumption and Heavy Pollution Industrial Processes.

The production of raw materials based on processing and utilization of mineral resources is the largest-scale production process, with intensive material and energy flow and great burden on environment. China's per capita high-quality mineral resources is in shortage, while the processing of existing resources mainly applies traditional and backward technologies. A large number of toxic and harmful materials are used, with low energy utilization efficiency and heavy environmental pollution. It is urgently needed to develop clean and comprehensive application technology system of circular economy with high efficiency which can significantly increase utilization rate of resources and greatly reduce burdens to environment. The frontier research areas include: ① to develop atom-economic new clean technologies for mineral resources transformation, establish new green process of substitutes for toxic raw materials, with high-efficiency and energy saving and material multi-level cycling process, Optimize processes and facilities from different scales for low environmental load design, amplification, integration and control; ② to establish in-depth comprehensive utilization of resources in multi-component green separation and recycling methods and eco-industrial network with zero emission; ③ to develop integration technologies of deep processing of mineral resources and high value products. For example, we need to make breakthrough in China's Vanadium-titanium magnetite and other special strategic metal minerals moderate and high-efficient transformation new-media technology, multi-metallic green separation new theories and high-value products recycling utilization new system; critical basic processes industrial clean production alternative technologies; such as atom economic reaction and green process engineering new technologies of phosgene, hydrocyanic acid and other toxic materials's (media's) important organic chemical intermediates production.

Atom-economic reaction

Atom-economic reaction is the most ideal reaction type, in which, all atoms of raw materials in production are 100% turned to products, while not to be turned to byproducts or waste. The measurement standard is atom economy (atom efficiency). The concept of atom economy was initiated by B.M. Trost, a professor of Stanford University, the United States in 1991. The definition of atom economy is the percentage of objective products to total reactants. That is Atom efficiency = (molecular weight of desired products / molecular weight of all reactants) × 100%.

In recent years, the development of atom economic reaction has become a hot spot of green chemistry. For example, in 1997, BCH company developed a new synthesis technology of ibuprofen (Ibuprofen is a widely used non-steroidal sedation, pain medication). The traditional production processes include 6 steps of chemistry-metric reaction, whose atom efficiency was less than 40%, while the new technology applies 3 steps of catalytic reaction, the atom efficiency reached to 80%. If taking into acetic acid by-product recycling, the atom efficiency can reach to 99%.

Institute of Process Engineering, the Chinese Academy of Sciences has successfully developed green chromium chemical clean production technology and integration, which replaces traditional high-temperature large kiln calcination reaction process with sub-molten salt clean and efficient homogeneous reaction process, significantly improved the recovery rate of chromium, and realized zero-emission clean production by prevention pollution from source in chromium chemical industry.

Source: Trost BM. 1991. The atom economy: a search for synthetic efficiency. Science, 254: 1471
Fuan Wang, Baozeng Ren. 2002. Introduction to green process engineering. Beijing: Chemical Industry Press

Mineral Resources

Rules for Implementation of *the Mineral Resources Law* stipulate that mineral resources refers to natural resources in solid, liquid, gaseous forms with utilization values and formed in geological processes. Rules for implementation of *the Mineral Resources Law* listed breakdown of mineral resources found in China, including common energy minerals, metallic minerals, non-metallic minerals and hydrosphere mineral, with a total of 168 categories of resources. Groundwater has a dual nature of mineral resources and water resources. Mineral resources include not only materials that can be utilized under current technological and economic conditions, but also materials with potential values which can be utilized in future conditions. The difference between mineral resources and biological resources is that the speed of regeneration of mineral resources is very slow or can not be regenerated. Thus, treasuring and protecting mineral resources are very important.

2. Recycling of Secondary Resources and Venous Industry

The traditional technology of resource processing only obtained 20%–30% components of raw materials, while the rest of 70%–80% components were emitted into environments in the form of waste. With the increasing shortage of primary resources, people started to extract useful resources from waste to use them for a second time. Therefore, waste residue, waste gas and other industrial waste emitted from production process and social waste generated from social consuming process have been secondary resources, including large industrial solid waste, electronic waste, municipal solid waste, discarded machinery and etc. Reuse of secondary resources will be an important source of resources in future. Till now, China's secondary resources utilization industry has developed rapidly, whose output has reached to 400 billion yuan, with a rapid growth speed of 30%. However, secondary resources utilization technology is still faced with problems of complex technology, low recovery rate of resources, heavy secondary environmental pollution, poor economic benefits and etc. In the next 20 to 50 years, we should resolve the problem of environmental effects in recovery and re-use process of secondary resources complex system, make a breakthrough for entire life-cycle assessment methods and product eco-design theory of recycling of waste resources, establish environmentally-friendly waste resources in-depth recovery and comprehensive utilization of new technology system, and form a sustainable manufacturing and consumption system with combination of "artery – vein".

Arterial Industry and Venous Industry

In accordance with the direction of material flow, the industry bearing resource – product – consumption industrial process is defined as arterial industry; Venous industry (resource recycling industry) is industry that applies advanced technologies to transform waste in process of production and consumption into resources and products which can be used again and realize reuse of all kinds of waste and resource recovery, for purposes of energy saving and environmental protection, under the premise of ensuring environment security, including two processes of transforming waste into recycling resources and recycling resources into products. Circular economy is an integrated material flow system consisting of arterial industry and venous industry.

Source: State Environmental Protection Administration. 2006. Standards for venous industry eco-industrial park (for trial implementation), HJ/T275-2006
Zhou Hongchun, et al. 2008. Circular economics. Beijing: China Development Press

In the near future, we need to make breakthroughs in the following technologies: ① short-range separation of polymetal and utilization of multi-component and environmental control technology of alumina red mud, iron and steel slag, zinc leaching slag, dust, zinc-containing soot and other large industrial waste hard to be disposed, and form an industry; ② to make a breakthrough in technologies of multi-component in-depth separation of metal or non-metal and recycling clean technologies of circuit boards or other electronic waste, scrap, copper waste and other dismantled waste, and form an industry; ③ for industrial waste gas, acetylene tail gas, coke gas and other large industrial carbon emissions, to develop technologies of low-cost recycling of valuable components, high-value conversion and eco-industrial chain links, and form a multi-industrial coupling mode; ④ for waste biological materials, social waste plastics, urban life sludge, to establish grading extraction of carbon, nitrogen, chlorine, sulfur, phosphor and other resources and use high-value technologies to achieve cross-industry integration.

3. Efficiently Use of Alternative Renewable Resources and Cycle-type Social Engineering

With depletion of conventional mineral resources, the development of a new generation of renewable alternative resources has become an important guarantee for global sustainable development, such as renewable energy, biomass resources, carbon dioxide and etc. However, the high-value recycling of biomass, carbon dioxide and other new resources on a large scale needs support and guide of technological innovation. It is expected that in the period from 2030 to 2050, China will achieve large scale high-efficient clean utilization of renewable alternative resources and China will enter into a material cycle-type society. The cutting-edge of technology and breakthroughs are as below: ① clean utilization technology of natural renewable resources. In immediate term, upgrading of clean of traditional heavy polluting industries of biomass utilization, such as industries of paper, leather, textile, fermentation and etc, has become the focus of clean technologies and circular economy research. We should significantly reduce the total amount of pollutants through innovation of key technologies. For example, research and develop clean technologies of utilization with high value of natural cellulose of replacing strong acid and strong alkali medium. In long term, we should make a breakthrough in technologies of biomass new media, new catalytic highly-efficient conversion and realize conversion of high-efficient bio-refinery of natural renewable resources and additional-value products. ② technologies of turning carbon dioxide and other low-carbon resources into resources and inverse carbon manufacturing. In immediate term, the research hotspots of carbon dioxide emission reduction and resource transformation are realization of low-energy synchronous utilization of carbon and oxygen resources of carbon dioxide, including large-scale synchronous utilization technology of flue gas carbon dioxide of power plants and methane gas of coalfield, clean technologies and engineering research of large organic chemical products with carbon dioxide as their raw material, resource recycling technologies and system integration of fixing carbon dioxide through carbonated solid waste; In long term, we will mainly develop inverse carbon technologies using solar energy and nuclear energy to directly capture ultra-low concentrations of carbon dioxide from the atmosphere and further transform it into hydrocarbon fuels and compounds.

4. Regional Circular Economy System Integration of Production, Consumption and Natural Ecosystems in Harmony

The objective of development of circular economy is to gradually build a circular-type society with resources utilized efficiently, clean and recycled, and to realize high degree of harmony between industrial production, social consumption process and natural ecosystems. At a large regional scale, we should study mechanisms of elements metabolism and material external cycle in the process of resource cycling utilization and develop industrial symbiotic link technologies, establish optimization integration methods and evaluation system of material flow, energy flow and information flow in resources utilization. At large inter-enterprise system scale, we should realize connections between material flow, energy flow and information flow, increase utilization efficiency of resources and energies in systems, and provide technological support and examples to the establishment of resource comprehensive recycling utilization.

In immediate term, we should make a breakthrough in technologies and mode of recycling economy construction in high-energy consumption and heavy pollution industries intensive regions and China's special resource processing regions to support regions' sustainable development. In the medium and long term, we should develop design and evaluation methods of ecological industrial chain, based on multi-objective optimization of environmental economy and product life-cycle, establish circular economy regional mode with broader resource cross-processing areas and larger regional-scale, and develop scientific and quantitative evaluation methods of clean production and circular economy and national technology standard system to provide support and guide for ultimate construction of a recycling society with material recycling utilization.

The roadmap of science and technology development in this field and the main scientific issues and key technologies are shown in Fig. 7.7 and Table 7.6.

Industrial Ecology, Product Lifecycle and Ecological Design

Industrial ecology applies methods of industrial ecology metabolism and material flow to study metabolism mechanism of industrial eco-systems and control methods, which requires no harm to environment and ecosystems in the whole lifecycle of products including processes of mining of raw materials, production, manufacturing, utilization and disposal of used products. At the end of 1980's, R.A. Frosch, an American environmental ecologist, simulated biological metabolic process and proposed concept of "industrial metabolism" for the first time. In 2000, International Association of Industrial Ecology was founded. An international conference has been held every two years since 2001. Its objective is to find innovative solutions to complex ecological problems. Since 1990s, eco-industrial park, as an important practice of industrial ecology, has been developed rapidly and been the theme of the world's industrial park area. The typical eco-industrial park is Kalunborg Eco-industrial Park in Denmark.

Product life cycle refers to the product life chain consisting of various links such as collection and processing of raw materials, product production, transportation, sales, use, maintenance, recycling and final disposal of products. Product Life Cycle Assessment (LCA) is a comprehensive assessment of input, output and the potential environmental impacts of product life cycle of a product system. LCA was originated from a series of analysis and evaluation of packaged products carried out in the United States from the late 1960s to the early 1970s, whose symbol is the environmental impact analysis and evaluation of beverage packaging bottle conducted by American Coca-Cola Company in 1969. In the early 1980s, LCA was active with the increasingly serious environmental problems.

Product eco-design refers to making products with advanced technology, good environmental coordination and reasonable economical efficiency, by using technological information, coordinate environmental information and economic information related to product in product lifecycle, and applying a kind of advanced systematic design method which uses concurrent design and other advanced theories.

Source: Graedel T E, Allenby B R. Industrial Ecology (second edition). NJ: Prentice Hall
Brezet H, van Hemel C. 1997. EcoDesign: A Promising Approach to Sustainable Production and Consumption. UNEP, France

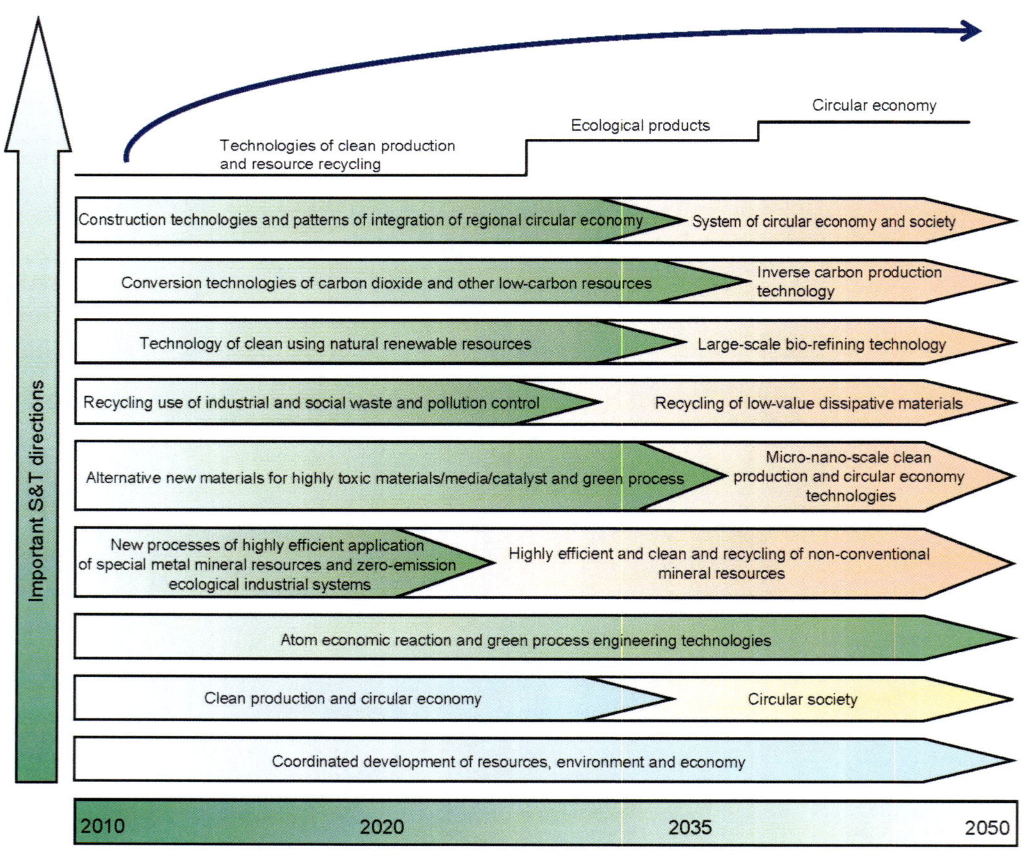

Fig. 7.7 China's S&T roadmap for clean production and circular economy

Table 7.6 Main scientific problems and key technologies in clean production and circular economy field

Main Objectives	Scientific and technological areas	Main scientific problems and key technologies
Control of pollution source and clean production	Environmental science and green process engineering	Design and realization of atom-economic reaction of minerals and fossil resources processing and transformation. Research and development of atom-economic reaction and high-selective reaction is in the core position of clean production technology and an important means of realizing "zero-emission" of process industrial waste. There are applicable examples in basic organic synthesis and chemical industries. Using non-toxic unharmful raw materials, solvents, catalysts in petro-chemical works has made progress in engineering. Original research results have been made in mineral resources cleaner production technologies. Now progresses of bio-technology, nanotechnology and other new technologies are making new breakthroughs. High-efficient, clean and recycling use of resources in green process engineering innovation. Including applying ionic liquids, supercritical fluids, new high-efficient catalytic materials and etc., to improve conversion efficiency of resources; increase process efficiency through micro-reactor, external field strengthening reactor, reaction-separation coupling and other processes to intensify making of equipment and methods to improve process efficiency
Resource recycling utilization	Environmental science and industrial ecology	High-efficient recycling utilization of alternative resources, including non-conventional resources (marine, extreme conditions, etc.), high-efficient recycling utilization of secondary resources, biomass resources, carbon dioxide and other low-carbon resources; Product ecological design and green recycling; a combination of "artery – vein," sustainable manufacturing system and ecological supply chain and etc. Multi-processes of material flow – energy flow – information flow environment – economy overall optimization and system integration method, production – consumption – natural large system ecological construction theories and technology, including environment – economy multi-objectives comprehensive evaluation indicator system, assessment method based on analysis of product life-cycle, complexity analysis of element metabolism chain and network between multi-systems and self-evolution design

7.7 Environmental Pollution and Health Effects

With the rapid socio-economic development and the continuous improvement of people's living standards, the categories and amount of pollutants entering into environment have kept increasing. In the next 40 to 50 years, it will be very difficult to effectively control toxic pollutants emissions to environments. History has proved that the continuous accumulation of toxic chemicals in environments will eventually do harm to human and ecosystems. The latest studies show that economic burden caused by air and water pollution accounts for a large proportion of GDP. Therefore, problem of environmental pollution and health effects is one of the fundamental problems which will affect China's future sustainable development. How to study impacts of pollution on human and ecosystems on molecular, cellular, individual, community and ecosystem levels is one of the most important objectives of environmental research, because the purpose of human's protection of environments is to protect the integrity of ecosystems and human health. Research on environmental pollution and health effects should not only focus on current conventional pollutants which human is faced with, but also focus on some new emerging pollutants such as antibiotics, internal secretion disruptors, algae toxins, oxidation disinfection by-products and etc. People are not clear about the environmental behaviors, exposure pathways and health effects of these new pollutants. The accumulation of new pollutants may bring greater risks and unknown longer potential impacts to ecosystems and human health.

If there is no systematic toxic mechanism and diagnostic techniques of environmental pollution to better analyze the health effects of pollution, we will not know which pollutants should be controlled, which links to control, nor can we develop effective pollution control technology, as well as develop effective management strategies to reduce these effects. Therefore, the content of environmental pollution and health effects research include the following aspects.

1. Toxic Mechanisms of Toxic Pollutants

We take typical model of animals or cell lines as our research object, apply technologies of genomics and proteomics to study toxic mechanisms of toxic pollutants, and reveal interaction sites of typical pollutants and life species and their roles at gene, cell and individual life levels. Metabonomics and other means are applied to study migration and transformation of pollutants in biological bodies and other metabolic ways, and its retention mechanisms in different biological tissues, and analyze mechanisms of impacts of toxic pollutants on biological metabolic processes and its control means. We will systematic analyze toxic mechanism of pollutants, at the same time, pay attention to toxic effects when different types of toxic pollutants coexist, reveal patterns and mechanisms of synergistic or antagonistic effects of mixed pollutants and quantitatively describe joint toxic effects of mixed pollutants.

2. Evaluation and Forecasting Techniques of Ecological Effects of Environmental Pollution

For main ecosystems, we should conduct sensitivity screening of different biological receptors, especially key species, and determine ecological relevant species as the main research objects. For typical species, study toxic mechanisms of environmental pollution at molecular and cellular levels, and study effects of environmental pollution at individual level. At the same time, like studies of health effects, we should systematically study toxic effects and joint toxicological mechanisms of mixed pollutants. At individual level, by building an eco-system model, we study patterns and mechanisms of interaction between toxic pollutants and organism, and develop a mathematical model of community ecotoxicology, based on simulation of ecosystem mechanism research. We will conduct long term fixed-point monitoring of some pollutants based on simulation research in laboratory, further study ecological effects of mixed pollution and mechanisms of community ecotoxicology, constantly improve the community toxicology and mixed pollution toxicology-based ecotoxicology mathematical model. Eventually, through integrated pollution effects, pollutant emissions and environmental behaviors of pollutants and etc., systematic model analysis is conducted to forecast and predict effects of toxic pollutants.

3. Rapid Diagnostic Techniques

The toxic effects of pollutants in environments are determined by the toxicity of pollutants themselves and the mechanisms of interaction between them and environments (that is, effective concentrations of pollutants in environments). Thus, rapid diagnostic techniques of environmental pollution effects need to develop *in situ* characterization techniques, environmental sampling techniques, sample pre-treatment techniques and rapid analysis techniques to examine the existence state of pollutants of environmental samples first. In addition, there are tens of thousands of toxic pollutants in environments and it is impossible to conduct toxicological tests for every kind of organisms. Therefore, it is necessary to develop *in vitro* testing technology for different categories and on different stages, based on research on patterns and sites of interaction between different types of pollutants. *In vitro* biological testing technologies and chemical analysis technologies will be effectively integrated to form rapid environmental pollutant recognition technology system. Based on studying and developing a toxicity recognition system, we should develop toxic pollutant rapid and on-line diagnostic techniques, form an environmental pollutant automatic recognition system with independent intellectual property rights, and promote roles of environmental pollutant automatic identification system in environment and health research and environment management through industrialization.

4. National Environmental Toxicant Screening Sharing Platform

There are huge regional differences in China's ecological environment and a wide variety of pollution. If there is no standardized toxicant screening platform, the research and management work of coordination of environments and health can not be fulfilled at the national level. In order to establish such a national sharing platform, first of all, an *in vivo* testing system of typical pollutants with Chinese characteristics needs to be developed. Through integration of toxicity identification technologies, *in vitro* testing technologies and online technologies of environmental samples toxicant screening technologies to form a platform, and then through systematic analysis of *in vitro* toxicity of typical toxicants, and by using toxic effects analysis of *in vivo* biological experiments, we should build a transformation mode for typical pollutants *in vitro* and *in vivo* testing, based on which, develop a calculation toxicology research approach system, and establish China's chemical products calculation toxicology and national toxicology database. Based on above research and technologies, China environmental sample toxicant screening technology system and technological platform will be established.

5. Environmental Disease Prediction and Forecasting Technologies

By systematically quantitatively measuring health risks of people in different regions exposed to various kinds of pollutants through the atmosphere, water, foods and other exposure ways, we will form environmental pollution exposure framework and exposure intensity of China's regional populations. Through analysis of specific biomarkers and population's pollution load, we will develop environmental epidemiology research approach system and regional population environmental epidemiology database with Chinese characteristics. Combining established exposure framework with epidemiological data, we will quantitatively analyze the quantitative relationship between pollutant exposure and human health, develop an evaluation model of China's environmental pollution and health effects, through modern information technology, form environmental disease prediction and forecasting technology system.

The roadmap of science and technology development in this field and the main scientific issues and key technologies are shown in Fig. 7.8 and Table 7.7.

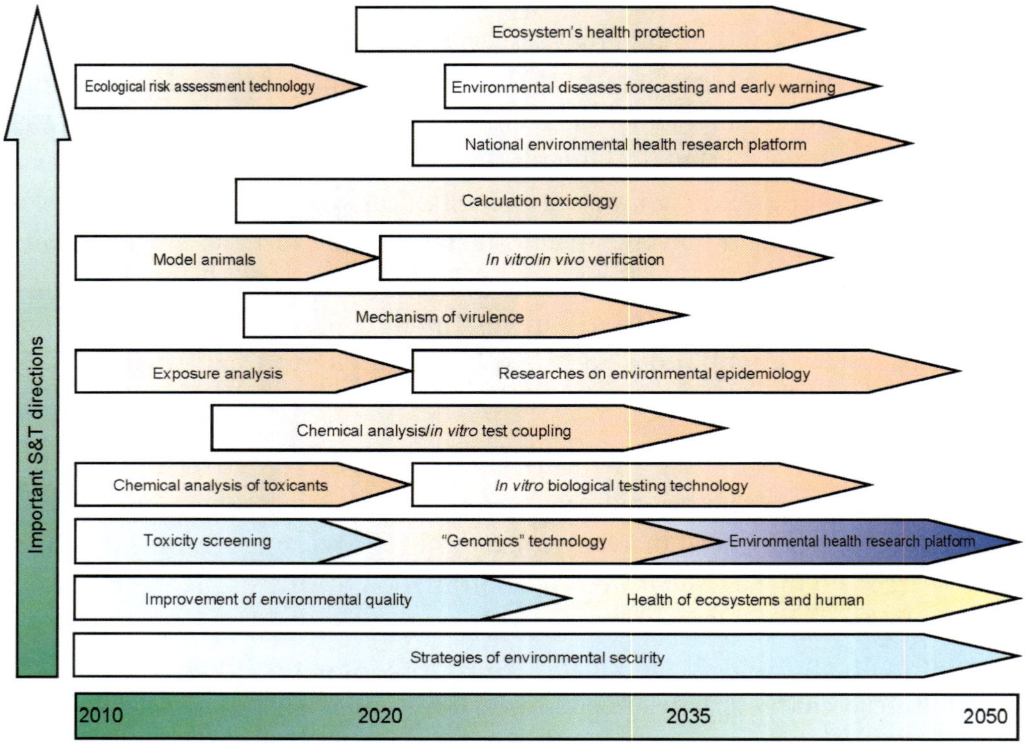

Fig. 7.8 China's S&T roadmap for environmental pollution and health effects

Table 7.7 Main scientific problems and key technologies in field of environmental pollution and health

Main Objectives	Scientific and Technological areas	Main scientific problems and key technologies
Mechanisms of toxicity of toxic pollutant	Environmental toxicology	Use genomics and proteomics technologies and etc. to reveal the mechanism of toxicity of toxic pollutants Use metabolomics analysis and other means to analyze mechanism of impacts of toxic pollutants on biological metabolic processes and its control means Analyze toxicity mechanism of mixed pollution and its characteristics of toxic effects
Ecological effects assessment and forecasting techniques	Environmental science, ecology	Apply functional genomics to reveal the mechanism of ecotoxicology of toxic pollutants Mechanism of community toxicology based on ecological complexity, analyze long term impacts of environmental pollution on ecosystems Forecasting and prediction technologies of ecological effects of toxic pollutants
Rapid diagnosis technical system	Environmental science, biotechnology	Technology of chemical characterization of pollutants in environmental samples *In vitro* testing technologies based on toxicity mechanisms Toxicity identification technology system based on chemical characterization and *in vitro* biological testing Toxic pollutants rapid and online diagnostic techniques based on toxicity identification technology system
National Toxicology Screening sharing platform	Zoology, biotechnology, bioinformatics	Typical pollutants *in vivo* testing system with Chinese significance Integrate toxicity identification technologies, *in vitro* testing technologies and reproduction technologies to establish an environmental samples toxicant screening technology platform Establish calculation toxicology research and method system to form a database of toxic chemicals in China
Prediction and forecasting of environmental diseases	Environmental science, environmental epidemiology, mathematics	Quantitatively measure exposure of populations to water pollutants and their health risks Quantitatively analyze exposure of populations to food pollutants, establish regional scale assessment system of health risks of food pollutants Quantitatively measure health risks of air pollutants, establish regional scale of air pollution health risk evaluation system Analyze quantitative relationships between environmental epidemic diseases and pollution exposure and human pollution load Pollutant exposure scientific system suitable to China's populations, combining with modern information technology to form an environment diseases prediction and forecasting technology system

Research on Protection of Environmental Health Transition

With the rapid socio-economic development, tens of thousands of new chemicals have entered into the environment and have affected human health through various exposure ways. Traditional animal body test-based study apparently can not keep up with the pace of an increasing number of chemicals. In addition, such kind of *in vivo* experiments highly relying on higher animals are usually time-consuming, and subject to a number of management regulations. Therefore, in 2008, Collins, Director of Institute of National Human Genome of the United States, pointed out that we should make a transition from past mainly relying upon animal body *in vivo* experiments to *in vitro* experiments and lower animals *in vitro* experiments and calculation toxicology research in "Science".

Source: Collins F S, Gray G M, Bucher J R. 2008. Transforming environmental health protection. Science, 319: 906–907

U.S. National Toxicology Program for the 21st Century – the Development Roadmap for Strategic Objectives

NTP (National Toxicology Program) is a national toxicology program which researches and applies modern toxicology and molecular biology tools to evaluate relevant public health and hygiene, and safety of chemical substances. It was established in 1978, and conducted complex toxicology testing research in aspect of improving evaluation of impacts of chemical and physical exposure on human health and the potential effect and solved many complex interdisciplinary key problems. Now it is one of the world's leading organizations in terms of providing and improving chemical substances' potential security.

1. Challenges Faced by U.S. National Toxicology Program

Through traditional toxicity testing methods, U.S. National Toxicology Program designed and conducted many animal experiments. However, traditional toxicity testing method often requires using large batches of tested animals and is time-consuming. Now, about eighty thousands of chemicals are used in businesses, among which, only a small number of chemicals have passed toxicity testing evaluation. Moreover, more and more new synthetic chemicals keep emerging. U.S. Environmental Protection Agency receives more than 2,300 kinds of synthetic materials applications. On the one hand, about the safety of human exposure to chemicals is still lack of scientific data; on the other hand, the complexity of environmental exposure and the uncertainty of evaluation of risks at different ages also make U.S. National Toxicology Program, Department of Health, and research institutions faced with enormous challenges in explaining toxic effects.

2. Changes of U.S. National Toxicology Program on environmental health protection

U.S. Environmental Protection Agency and U.S. National Toxicology Program funded by U.S. National Research Council developed a long term plan for strategic objectives of toxicity testing and a strategic plan to achieve this long-term goal. In 2004, U.S. National Toxicology Program announced the strategic objectives of the 21st century and the roadmap of development, whose mark is a toxicity testing method combined with innovative high-throughput screening and other automated screening analysis. In 2005, U.S. Environmental Protection Agency established National Calculation Toxicology Research Center. Through these innovative activities, U.S. National Toxicology Program, U.S. Environmental Protection Agency and

Chemistry – Gene Research Center of U.S. National Institutes of Health Sciences made the strategic development objectives for the 21st century, i.e, promoting the transformation of toxicology from an observational science mainly based on models of disease characteristics of internal body to a predictable science mainly based on mechanism of observation of biological target characteristics *in vitro*.

3. Ways to implement the strategic objectives in the 21st Century of U.S. National Toxicology Program

(1) High-throughput method U.S. National Toxicology Program plans to apply the up-to-date physiological, biochemical and molecular knowledge to study high-throughput method and forecast toxicity effects of environmental incidents in the next ten years and to confirm that only through innovative application of assessment tools based on mechanism of toxicity, can it provides a huge amount of necessary scientific information to decision-makers for public health, reducing the burden caused by environmental diseases.

In vivo and *in vitro* tests are often used as identification tools of cell response after being exposed to chemical substances to obtain toxicity response effects. U.S. Environmental Protection Agency, U.S. National Toxicology Program and Chemistry – Gene Research Center of U.S. National Institutes of Health Sciences are developing a new high-throughput method, whose test range is 5 nmol – 100 μmol, which can produce a concentration – effect curve. This method can be repeated for many times. Compared with traditional analysis methods, this new method has advantages of good repetition. The developed information platform can compare the results of high-throughput identification. This comparison can be extended to U.S. Toxicology Program and U.S. Environmental Protection Agency previous data. Moreover, all collected high-throughput data can be obtained via web access.

(2)Study on toxicity of chemical substances and its testing At present, Chemistry – Gene Research Center of U.S. National Institutes of Health Sciences is conducting biochemical analysis of 50 kinds of cells of more than 1,800 kinds of compounds provide U.S. National Toxicology Program and U.S. Environmental Protection Agency. The result of analysis shows that 1,408 kinds of compounds from U.S. National Toxicology Program can produce cellular toxicity to human and 13 kinds of rodents. Some compounds can produce toxicity to all cell populations, while others are selective. This indicates that titration-based high-throughput identification analysis can produce high quality *in vitro* toxicity data for thousands of substances, and the complexity of selecting appropriate cell types and analysis ends.

Another method to identify toxicity way is to explore the diversity of responses of human and animals to known toxicants. Chemistry – Gene Research Center of U.S. National Institutes of Health Sciences is evaluating sensitivity of human cells to 2,800 kinds of compounds provided by U.S. National Toxicology Program and U.S Environmental Protection Agency. Similarly, U.S. National Toxicology Program has established a test method which uses different mice to test their different responses to diseases. Chemistry – Gene Research Center of U.S. National Institutes of Health Sciences has also evaluated the sensitivity of animal cells to compounds. The collected data results are used to study the ability of biological activity, which can be used to predict observed types of quasi-toxicity tests to identify biological ways leading to toxic effects. Its ultimate goal is to build relationship between mice and external signals of human body *in vivo* toxicity tests. In addition, it is developing a calculation method in order to understand biological outcome of exposure and simulate biological responses of specific organs to exposure.

(3) From mammals *in vivo* test to *in vitro* non-mammals animals In the past 20 years, U.S. National Toxicology Program expanded the evaluation range of individual cases and biological evaluation of final numbers of species at the ending points, especially for biennium cancer research on rats' slow exposure. At the beginning stage of realizing the roadmap will involve evaluation of all analyzing methods to analyze their forecasting capabilities and confirm whether these research modes need changes. These comments include selection of types of wild populations, model changes, life stage when exposed, sufficient response endpoints and other endpoints in order to obtain the best decision-making objectives by applying sciences.

Time limit: Within 12 months, U.S. National Toxicology Program will develop a strategy, that is, design genetic analysis of targets of response endpoints for current research which can be operated everyday. The objective of this analysis is to provide optimal selection and mechanism information for improving understanding of specific chemical substances toxicological information and the target of mechanism of high-throughput analysis, the results of this study can also be applied to other analysis of mechanism of chemical substances. These key factors of the analysis include identification of target genes and gene products, platform selection, and databases of National Environment and Health Research Center – National Toxicogenomics Research Center and members of U.S. National Toxicology Program and connections of results of analysis.

Time limit: Within 18 months, U.S. National Toxicology Program plans to check the designs of mice biological testing experiments conducted in the past two years; within three years, will comment on designs of cancer researches (including developmental toxicology, immunological toxicology, neurological toxicology, etc.) to ensure that the generated data can identify cases of human health effects which have caught great concerns. The strategic objective of U.S. National Toxicology Program is to study the metabolism of exposed animals on various stages and to promote the application of gene and protein expression technologies, which will help to describe the toxicity characteristics of substances and understand the mechanisms of the occurrence of diseases and their deterioration caused by toxicants and identify useful biomarkers of disease with exposure to toxic substances.

Time limits: in the two years of mice cancer research of U.S. National Toxicology Program, a pressing goal is to capture digital images from toxicological slices everyday, and develop technologies of obtaining systematic evaluation of damages and tumors from analysis of images. The long term goal is to provide digital image files for present and past researches of U.S. National Toxicology Program. Finally, tools that obtain images from tissue slices, dead organisms and live animals will be significantly improved in the last five years. U.S. National Toxicology Program will apply these technologies as soon as possible to biological analysis, to improve the speed and accuracy of comments on toxicity, and to make modes of comments change and tissue slices include two-dimensional and three-dimensional measurement. These image tools can significantly improve the statistical effectiveness of researches of U.S. National Toxicology Program and reduce the number of required animals for specific analysis.

Time limit: U.S. National Toxicology Program plans to build a project team to explore an effective non-mammalian model in the next 3 years, at the same time, continue to explore non-mammalian tests *in vivo* analysis as the alternative model of toxicity tests. The project of "Caenorhabditis elegans" of U.S. National Toxicology Program will be completed within the next 3 years. The purpose of this project is to use "Caenorhabditis elegans" as an effective analytical tool for developmental, neural and behavioral toxicological researches. If "Caenorhabditis elegans" have sufficient sensitivity to specific reagents, U.S. National Toxicology Program will expand this project and apply it to more chemicals. After the completion of this project, U.S. National Toxicology Program will evaluate other toxicity endpoints of other non-mammalian species as a potential high-throughput analytical method.

Sources: National Institutes of Health, National Institute of Environmental Sciences. 2004. A National Toxicology Program for the 21st Century

7.8 Advanced Monitoring and Forecasting Techniques

Research on ecology and environment can not be separated from advanced monitoring techniques. Monitoring techniques can strengthen researches on modern sampling, sample preparation techniques and modern environmental monitoring analytical techniques for different processes of physics, chemistry and biology, especially, focusing on strengthening environmental sensors, ultra-trace monitoring and analysis technologies, on-line monitoring techniques, morphological analysis technologies, and source apportionment technology studies. At the same time, we conduct monitoring on molecular, cellular, individual, community, ecosystem, regional and even global scales. Moreover, there are different requirements in terms of time series, such as online, long-term positioning, real-time transmission and etc. to achieve scaling conversion, systematic integration and assembly of technologies, and on this basis to carry out environmental monitoring network optimization studies. Researches on ecology and environment, not only need to clarify the processes of the past and the present, but also need to understand laws through these studies, develop system simulation and emulation technologies for environmental processes, and applied researches on pollution accident monitoring equipment and early warning and forecasting techniques. The main development directions include the following aspects.

1. Advanced Chemical and Biological Sensors and Their Studies and Integration Application in Online Environmental Monitoring Technology

Researches on ecology and environment are turning from short-term observation to long-term positioning observation studies. Moreover, molecular biology techniques are increasingly being infiltrated into the ecological and environmental studies, that is, from the studies of ecological and environmental effects of emission of pollutants, to studies of micro-impact of pollutants on health. This undoubtedly sets higher requirements for environmental monitoring technology, that is, it is needed to conduct real-time continuous online monitoring of various types of environmental factors and pollutants, so as to promote sensor technology to develop toward embedded, integration, automation, intelligent direction and networking directions. The focus of the studies include: ① to strengthen the applications of high-tech, such as nanotechnology, supramolecular chemistry, surface self-assembly and gene technology in modification of sensitive layer of surface of sensors; ② applications of technologies of miniaturization and micro-electro-mechanical systems in micro-sensors and integrated system processing; ③ research and development of environmental embedded sensors, multi-sensor integrated system; ④ research and development of chemical-biological sensors based on bionic technologies; ⑤ research and development of testing data long-distance wireless transmission technology.

2. Building of Wireless Data Transmission – Terminal Handling – Model (simulation coupling system) and the Formation of Pollution early Warning and Forecasting Techniques

Building Wireless data transmission – terminal handling – model (simulation coupling system) and forming pollution early warning and forecasting techniques, are the urgent requirements of enhancing the technological level of China's environmental protection and promoting the building of an environment-friendly society.

The focuses of the studies include: ① to develop an early warning technology system with technologies of sensors, simulation models, information processing technology as the core of it, and build an early warning technology platform of grave environmental pollution incidents; ② to research and develop a multi-scale, multi-media, visual, dynamic environmental risk site simulation system; ③ to establish an information database of functional materials of rapid detection techniques for environmental pollution incidents, research and develop rapid spot detection techniques, key technologies and equipment for environmental accidents characteristics of pollutants and establish a spot rapid detection system for characteristics of pollutants.

3. Ground/air Observation Technology and the Digital Earth

The combination of ground observation technology and satellite remote sensing technology is not only the inevitable choice for the realization of dynamic monitoring, scientific assessment and prediction of regional ecosystems and environment changes, but also the main channel of systematic development of digital earth and obtain multi-sources and massive scientific data. Breaking fragmentation and industry barriers and building China's true data accessing and sharing system for the Earth's surface geographical, ecological and environmental factors regional distribution and their dynamic changes, with integration of geographic information system – terrestrial observing system – satellite remote sensing observation system, are the major problems that China must resolve currently. The focuses of the studies include: ① design of dynamic monitoring system of the earth's surface geographic, ecological and environmental factors and comprehensive application of the advanced technologies; ② integration and management technology of diversified, distributed and massive data; ③ land surface process model and analytical applications software, test data, observation data, remote sensing images and spatial database integration technology.

4. System Simulation of Environmental Processes and Simulation Technology

Comprehensively apply modern analytical techniques for environmental monitoring, mathematical model tools and computer technologies to perform system simulation and simulation of environmental processes, which is an important tool for understanding regional ecosystems and environmental changes and its mechanism, and is also the scientific basis of environment quality early warning and forecasting. The focuses of the studies include: ① Integration of advanced monitoring technologies of interface processes of multi-media environmental factors and their applications; ② development of simulation technology system based on environmental process models; ③ system simulation and simulation technologies of ecological and environmental change processes based on ground/air observation technology and digital earth system.

The roadmap of science and technology development in this field and the main scientific issues and key technologies are shown in Fig. 7.9 and Table 7.8

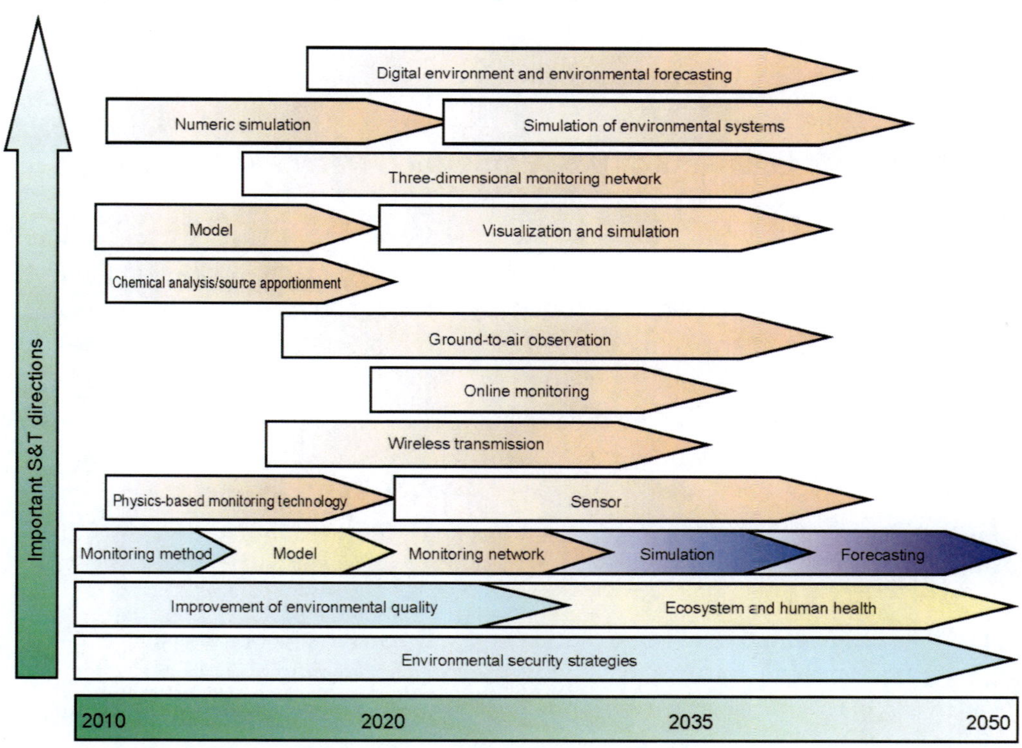

Fig. 7.9 China's S&T roadmap for advanced monitoring and forecasting technology

Table 7.8 Major scientific issues and key technologies of advanced monitoring and forecasting technologies

Main Objectives	Scientific and technological areas	Major scientific issues and key technologies
Environmental sensors	Environmental sciences, analytical chemistry	Research and application of advanced chemical and biological sensors for ecosystem functions and three environmental media of water, soil and air
Online monitoring technology	Environmental Sciences, physics, information technologies	Physics-based online environmental monitoring technology and its equipment making
Environmental monitoring network	Environmental Sciences, information technologies	Conduct monitoring on molecular, cellular, individual, community, ecosystem, regional and even global scales; also there are different requirements of online, long-term positioning, real-time transmission and etc. in time series; conduct environmental monitoring network optimal adjustment studies
Pollution early warning and forecasting technologies	Environmental Sciences, information technologies	Build wireless data transmission – Terminal handling – Model/Simulation coupled system; form pollution early warning and forecasting technologies
Ground/air observation technology and digital Earth	Environmental Sciences, information technologies	Establish and improve space-based and ground-based integrated (integration of satellite, machine, and the ground) Earth system and ecological environment monitoring technology and method system, resolve key technologies of time and space precision, standard-setting, multi-source data fusion and integration, key parameters accurate inversion and etc. of resources and environmental monitoring

Research Overview of U.S. Embedded Networked Sensing

Embedded networked sensing (ENS) system is a muster of a large number of intelligent sensors and actuators distributed in the physical world. It can make environmental monitoring intensive in time and space, and thus can enable people to observed phenomena more closely. It can also reveal phenomena which can not be observed before, and enable people to better understand and manage the complex natural environment. ENS needs to use wireless and long-term systems without people's care, and process data from data sources, i.e, near sensors, to screen useless data, so that energy consumption of communications can be reduced. ENS also needs self-configuring systems to adapt to unpredictable environments, and also needs network internal collaborative signal processing in order to achieve distributed control.

Center for embedded networked sensing (CENS) is one of the six technology centers sponsored by U.S. NSF, which was founded in 2002. It is also an interdisciplinary team. Its team members include thousands of teachers, engineers, graduate students and undergraduate students in multiple disciplines from University of California, Los Angeles, University of Southern California, University of California,

Riverside, California Institute of Technology, University of California, Merced and University of California, Berkeley. The goal of CENS is to research and develop ENS system, and apply it to important scientific researches and social applications.

Current ENS applied researches in CENS mainly include the following five aspects, namely, terrestrial ecosystem observation, monitoring and management of migration of pollutants, aquatic microbial observation, earthquakes and urban sensing. Here a brief introduction to three aspects of their applications in ecological environment area.

1. Terrestrial ecosystems observations (TEOS)

The main task is to design, develop, deploy, evaluate and maintain embedded network sensing systems to conduct continuous original *in situ* monitoring of environmental, physiological and ecological variables of various terrestrial ecosystems. Up to now, CENS has successfully set up many kinds of sensors systems, cameras and platforms in James preservation area field observation station. This preservation area provides a prototype deployment and test base for U.S. Ecological Observation Network (NEON). The research team has focused more on multi-scale and multi-modal methods of observation of ecosystem processes. With the increase of integration degree of digital cameras and the increase of their stability, as well as image analysis technology becoming more mature, using cameras as biological sensors has been placed on the agenda. The research team has also used low-cost consumer camera and web cameras connected to the Internet to conduct quantitative ecological study. The research of the research team include the following five areas:

(1) Animal observation camera system. The aim is to research and develop camera-based animal observation system to investigate behaviors of birds and reptiles. The required key technology is to increase the acquisition frequency of data, and research and develop an image processing software package which can automatically classify behaviors of birds. The main task is to deploy Cyclops camera system to meet the needs of research. Up to now, the two tower cameras, one of which was installed in station of UCNRS, observing wild animals, the other installed in JR , have automatically collected and recorded positions of nests of birds of prey and activities of bighorn sheep, and have continuously and reliably provided and uploaded images.

(2) Plant Observation Camera System. The fixed high-resolution digital cameras installed in James preservation area have participated in many application research projects as phenology changes and physiological state biological sensors. Time of flowering and leaves growth can be obtained very easily through quick digital images. And provide measurements of conditions of phenology, area changes of leaves and photosynthesis for analysis of greenness of leaves in visible light images and landscapes. Daily images of woody and herbaceous plants can detect changes in time of flowering and leaves. A combination of fern leaf change images and data of apparatus monitoring of changes in solar radiation and photosynthetic carbon assimilation, can provide a new method of understanding responses of stomas to light environment. At present, Moss-CAM digital camera is used to study seasonal changes of stellate moss photosynthetic response.

(3) Automated minirhizotron and arrayed rhizosphere soil sensors (AMARSS). Through modeling to study the relationship between micro-climate on the ground

and soil energy balance measurement and underground soil environment measurement, to study temporal and spatial anisotropy of soil. Work of measuring soil environment in James Nature Reserve has kept colleting high-resolution soil spatio-temporal data. 10 observation stations have been set up on an 80 m section in an understory leaf layer. Every station is observing the ground and underground area, trough sensors group. On the ground, mobile NIMS is applied to monitor temperature, relative humidity, PAR and other indicators, underground, AMARSS is applied to measure temperature, moisture, soil carbon dioxide and other environmental indicators.

(4) Environmental sensing and imaging accessing infrastructure systems. Environmental sensing system is the key part of a wireless sensor network. The latter provides end-to-end, sensor to user services. After running for more than five years, this study replaces CMS with environmental sensing system-based CMS2.

(5) Hand-held data collection system (Eco-PDA). It is a supplementary to embedded sensing system, through support of more space coverage areas and higher human observation.

2. Monitoring and management of pollutant migration

The effort is dedicated to research and develop ENS technology which supports matter and energy distribution and flux monitoring on different scales. The objects of research include soil, groundwater and waterfront systems. Since 2006–2007, the focus of research has transferred from laboratory and small-scale experiments to real environment on meaningful space-time scale. In the last 12 months, 13 times of field deployments were taken. Involved areas include water quality of underground water, observation of rivers and cities' rivers and etc.

(1) Water quality of groundwater. In Spring 2006, soil tower is used to conduct large scale experiments at Agricultural Experiment Station in Palmdale, the main content of the experiments are applying soil sensors pylon (Soil Pylon) to measure soil temperature, humidity and some interesting substances, such as nitrate and etc.; monitoring of water level dynamics, chemical concentration of groundwater by using pressure and chemical Javelin.

Soil tower is a vertical distributed one-dimensional sensor array put in soil, which is used to measure soil moisture, energy flow and fluxes of interesting chemical substances to characterize flows of matter and energy of underground system. The sensor array of soil tower includes temperature sensors, moisture sensors and a series of water quality sensors measuring PH value and concentrations of special ions (such as NO_3^-).

Pressure javelin consists of PVC pipe with length of 2.75 m, diameter of 1.25 inch. It is slotted at its bottom to allow water to enter into it, and installed with a pressure transducer to measure the water level of internal tube. Chemical javelin is similar to pressure javelin, however, it is equipped with ion-selective electrodes to detect NO_3^- and NH_4^+.

(2) Observations of rivers and urban watercourse. In a main river section, the task is to deploy a multi-scale ENS, install NIMS on a 70 m wide span, use various types of sensors to conduct raster scan to realize automatic measuring of velocity of flow, water quality and other indicators of cross-section, including temperature, salinity, dissolved oxygen, electric potential of redox potential and NO_3^-, NH_4^+.

3. Observation of aquatic micro-organisms

Using new wireless sensor network to *in situ* monitor time of aquatic organisms birth and death, verify assumptions of processes controlling distribution, growth and death, to simulate and predict the population dynamic trends. Through early laboratory design, in 2005 and 2006, the team built and deployed a distributed sensor system NAMOS in Lake Fulmo in James Natural Reserve, which can obtain a huge amount of information of continuous physical, chemical and biological characteristics of water bodies, and can conduct intensive monitoring of specific sites in important periods. Recently, two tests for coastal ecosystems have been done.

Source: Translation from http://research.cens.ucla.edu/

Wireless Sensor Network

Wireless sensor network is currently a hot research area with cross-multidisciplinary and highly intensive knowledge in international community, which has caught a lot of attentions. It integrates technologies of sensors, embedded computing, modern networking and wireless communications, distributed information processing and etc, which can conduct real time monitoring, perception and collect information of various types of environments or monitored objects, through collaboration of various types of integrated micro-sensors. This information is sent through wireless mode and transmitted to end users through multi-hop ad hoc network, to achieve the connections between physical world, computing world and human society. Wireless sensor networks has very broad application prospects and has potential application values in many important areas such as military defense, industries and agriculture, urban management, bio-medical, environmental monitoring, emergency rescue and disaster relief, counter-terrorism, hazardous area, remote control and etc., which has caught a lot of attentions from many countries' academia and industrial sector and been considered as one of the most important technologies that will cause great impacts on the 21st century.

A typical system architecture of wireless sensor network consists of distributed sensor nodes (Group), receiving transmitters, Internet and user interface and etc. Among them, wireless sensor network nodes consist of four basic modules: sensing unit (consists of sensors and analog-to-digital conversion module), processing units (including CPU, memory, embedded operating system, etc.), communication unit (consists of wireless communication modules), and power.

In addition, it can be selectively equipped with other functional units, such as, positioning systems, mobile systems, power self-supply systems, etc. In wireless sensor networks, nodes can be put in perceived objects or near them, through aircraft dispensing mode or artificial arrangement mode. These nodes constitute wireless networks through self-organizing mode, and perceive, collect and process information in areas covered by the networks in a collaborative way, and through multi-hop network, transmit data in the whole area through sink nodes (receiving transmitter) to the remote control and management center. Vice versa, the remote management center can also control and manipulate nodes of networks in real time.

Wireless sensor network node is a miniaturization and embedded system, which constitutes a support platform for wireless Sensor network. Up to now, there have been designs of many types of network nodes both at home and abroad. Their realization principles are similar, just using different microprocessor or different communication or agreement protocols respectively, for example, custom protocol, 802.11 Protocol, Zigbee Protocol, Bluetooth Protocol, and UWB communication mode. Typical nodes include Berkeley Motes, Senssoria WINS, Berkeley Piconodes, MIT μAMPs, SmartMesh Dust mote, Intel iMote as well as Intel XScale nodes, ICTCAS / HKUST's BUDS, etc.

Wireless sensor network has great application prospect. There are countless projects that are constructed on platforms with various types of sensor network nodes to meet needs of land, sea and air all-around applications. For example, it can be applied to environmental monitoring, meteorological observation and weather forecasting, micro-biome observation, flood warning, farm management, intelligent home, intelligent transportation, radiation monitoring study, for Cricket and Echo for positioning purpose, health care projects SSIM and etc. With in-depth study and the development of their applications, a variety of wireless sensor networks will eventually spread all over the people's living environment, and thus truly realize "ubiquitous computing."

Source: Li Cui, et al. 2005. The progress of research on wireless Sensor networks. Computer Research and Development, 42 (1): 163–174

The above-mentioned eight important areas are closely tied to the needs of socio-economic development of China to 2050, and combined with the disciplinary frontier of global ecological and environmental sciences and technologies development, and are set up for China's current environmental and ecological conditions. The areas are interconnected, which constitute the overall framework of China's ecological and environmental sciences and technologies in the next 50 years. The relationships between them are shown in Fig. 7.10.

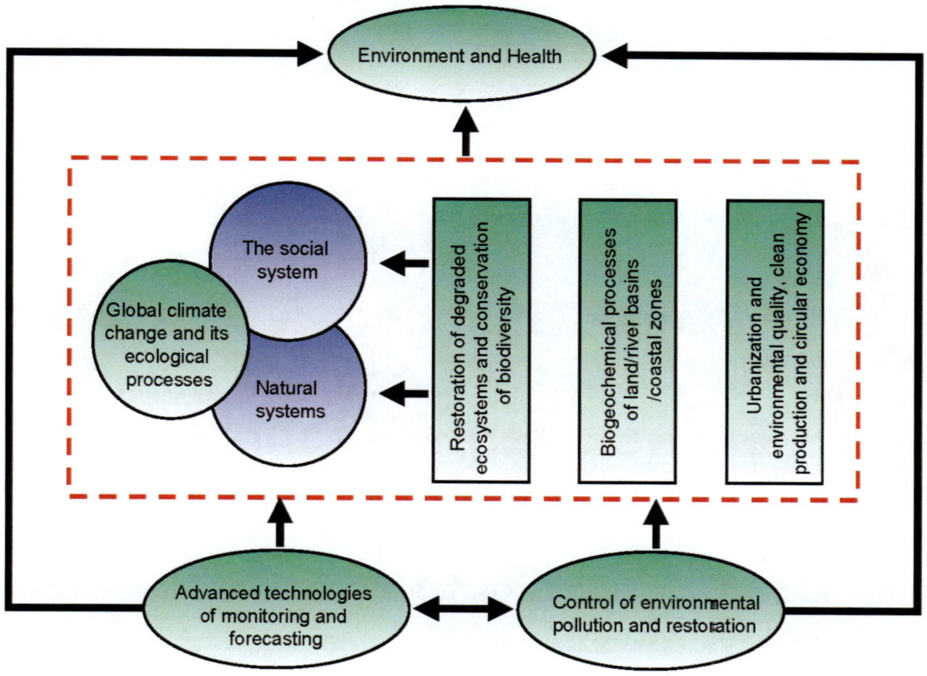

Fig. 7.10 Relationships between eight major areas

8 Generic Technology Roadmap for Areas of Ecological and Environmental Technologies

8.1 Multi-disciplinary Comprehensive Development of High-tech Technology

Modern scientific research has to rely on advanced scientific instruments and technological methods. The level of scientific research is increasingly dependent on the performance of scientific instruments and advanced technological methods. Currently, various disciplines keep on integrating and cross-cutting with each other, also keep differentiating from each other. Moreover, the integration time is earlier than before and the pace of integration is speeding up. Through multi-disciplinary cross, development of high-tech needs to make major breakthroughs in the following areas: sensor development based on physics, biology and chemistry; advanced chemistry and chemical engineering and materials science that promote green materials and processing technology and systems development; advanced rapid analysis techniques and technology of diagnosis of pollution; comprehensive applications of information technology and high-performance computing technology to enhance the levels of simulation, prediction and visualization technologies.

8.2 Ecological and Environmental Monitoring Platform and System Theory

To solve problems of irrigation, fertilization, carbon trading, biodiversity and waste management, we need clear and accurate data to provide a basis for environmental management decision-making. The ability of accurately measuring system dynamics is essential to environmental managers responding to dual challenges of environmental and economic sustainability. Building real-time, dynamic, visualized ecology and environment monitoring network can not only continuously manage resources, but also maintain competitive advantages in the increasingly competitive global market. Important theories and methods that need to be developed are as follows: development and deployment of embedded intelligent sensor network system; development of new methods and new theories applied to complex system study; building environment and health research general platform; forming comprehensive integration capabilities of different models and data, and developing forecasting techniques.

8.3 Data Integration and System Simulation

Qualitative data, data management and information systems are the prerequisite of reasonable environment scientific research and management. Collection and management of data needs to be improved, including data collection, data management, data accessibility and applications of new data collection technology. Integration of disciplines and transfer and absorption of research can not be realized without improvement of data management mode. Environmental managers are increasingly seeking information on how environments respond to changes and how to manage changes to achieve multiple goals from scientists, which needs to simulate complex systems, which can not only convey uncertain forecasting information, but also provide useful basis of decision making. These include: large-scale, long time series of environmental, socio-economic and health data acquisition; developing a series of numerical models used in scientific research, understanding the complexity and unpredictability; conducting large-scale simulation experiments and dynamically revealing and forecasting overall behaviors of complex systems, combined with models.

9 Overall Roadmap for Ecological and Environmental Technology Development

We should strive to build the science and technology support system of a harmonious society, by attaching great importance to innovation of basic theories, taking national strategies as our orientation and studying global changes and the characteristics of China's ecological systems from a global perspective. Through top-level designing, we should fix the position of China's ecological and environmental science and technology areas accurately, identify strategic focuses and develop the roadmap of science and technology development; develop theories and methods of observation, forecasting, simulation integration, decision-making, regulation and control of ecological and environmental studies on the whole, integrate them on space and time scales and lay the foundation for China's sustainable development. Through system integration, we should fully take advantage of China's advantages in ecological and environmental science and technology area to meet the requirements of national economic and social development and solve complex problems of ecological and environmental areas.

9.1 Objectives to 2020

To build a relatively sound ecological and environmental observation, monitoring, exploration, experimentation and experimental study platform, basically reaching international advanced level.

To resolve key technological problems of energy-saving and emission reduction technology, pollution control and remediation technology, monitoring technology and etc.

To make breakthroughs in building disaster emergency command systems and comprehensive information platform of risk management, biodiversity conservation, vulnerable ecosystem assessment, ecosystem management and complex pollution mechanism research, etc.

9.2 Objectives to 2030

To build a sound ecological and environmental observation, monitoring, detection, and testing and experimental research network, reaching the international advanced level.

To make breakthroughs in areas of major natural disasters risk investigation, risk level assessment techniques, rapid diagnostic techniques for pollutants, eco-factory integration technology, biodiversity conservation technology and alternative energy sources/materials, etc.

To understand the stability mechanisms of fragile ecosystems, take part in global research activities of construction of degraded ecosystems and international diplomacy and negotiation on construction of degraded ecosystems.

9.3 Objectives to 2050

To establish an ecological and environmental observation, monitoring, detection and simulation early warning system to predict and forecast changes in environment.

To resolve critical scientific and technological problems of protection of the integrity of ecosystems, forecasting environmental disease and substances cycles of ecosystems.

To basically explain the formation mechanisms and causing mechanisms of various disasters.

To make breakthroughs in technologies of acquisition of new energies, integrate and simulate multi-disciplinary, multi-scale and multi-variable complex systems, systematically understand behaviors of environmental systems, control and adjust global environmental changes and the process of socio-economic development, integrate ecological knowledge and information into decision-making and management, and promote the scientific guidance of the latest advancement of ecology to ecological sustainability in all-round.

9.4 Roadmap to 2050

The roadmap of China's ecological and environmental technology development to 2050 can be summarized from four aspects: response to global climate changes, river basin environment quality, urban environment quality, and biodiversity and ecosystem. The focus is to understand laws of evolution of environment on different scales of time and space, develop eco-system restoration and pollution control technology, establish three-dimensional monitoring network of evolution of ecological systems and quality of environment monitoring network, systematically arrange typical experimental model conservation areas, for solving core science problems, making breakthroughs in key technological problems, improving systems integration, promoting demonstration activities, etc. For details, see Fig. 9.1.

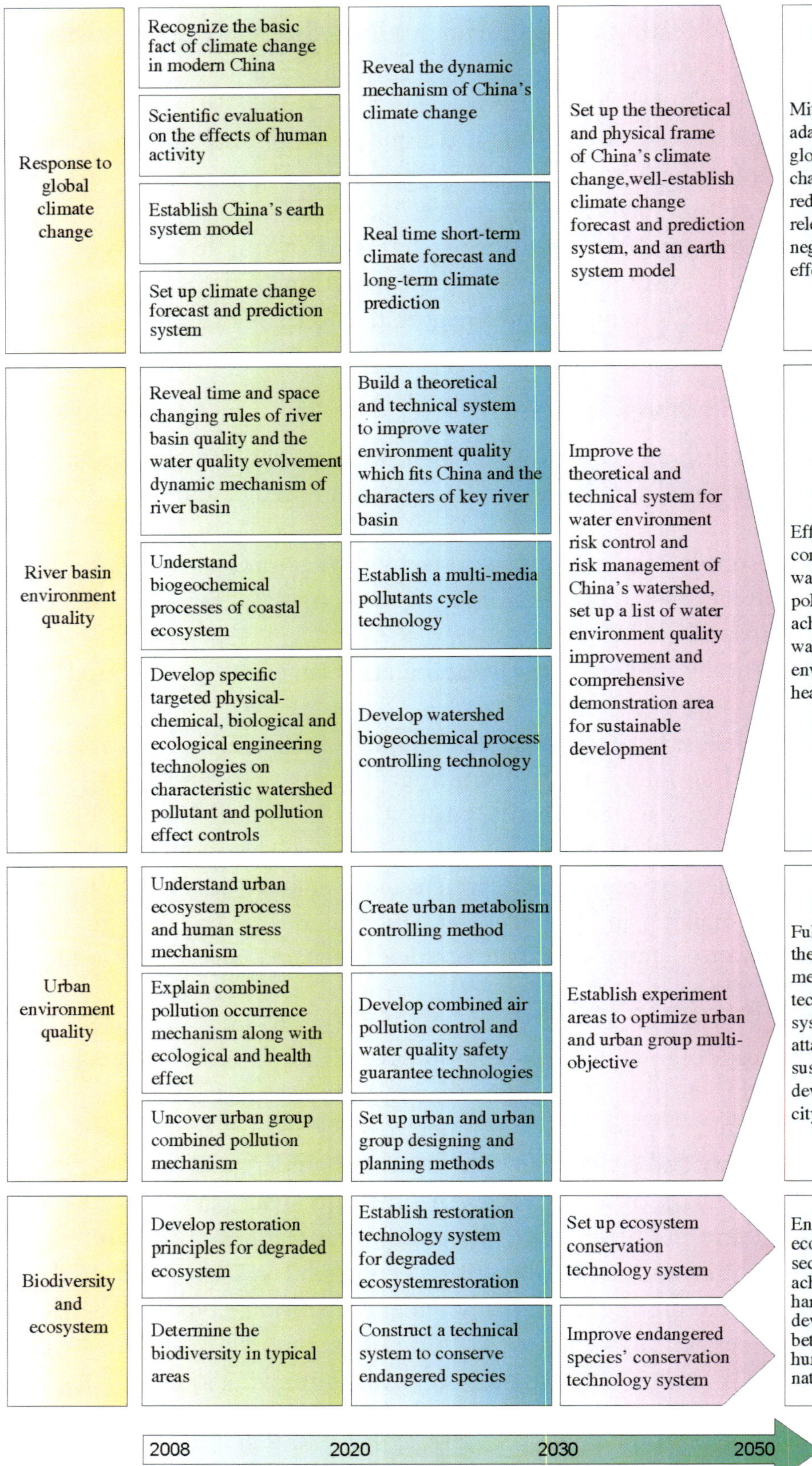

Fig. 9.1 China's S&T roadmap for ecological and environmental science to 2050

10 Security System and Implementation of Roadmap for Ecological and Environmental Technology Development

10.1 Launch Large Methodology-Targeted Research Projects

For research areas which need long-term and stable support, key research projects should be set up with methodology research as its objective. Through the research projects, we should enhance top-level design, co-ordinate the deployment of long-term, systematic basic data acquisition and basic process studies, normalize observation and test standards, strengthen the overall capacity of international cooperation, coordinate relevant research projects, take advantages of comprehensive multi-disciplines crossing and try to make theoretical breakthroughs in international hot-spot research areas.

10.2 Establish Problem-Oriented Interdisciplinary Ecological and Environmental Basic Research Platforms

Expediting construction of support systems and platforms; observation, exploration, investigation, testing and experimental research of modern technology is the basis and support of Earth sciences' research and development. We should establish problem-oriented relatively complete network of field observation experiment station for China's resources, ecosystems, environment and etc., realize long-term monitoring, observation, detection testing, early warning, forecasting and demonstration function of resources and ecological environment, and make full use of these platforms to carry out high-level research work.

10.3 Build 2 to 3 Comprehensive Experimental Demonstration Areas (cities) with Multiple Objectives Optimization under the Guidance of the Concept of Scientific Development

The concept of scientific development is the major strategic idea for China to develop socialism with Chinese characteristics, which China must adhere to and implement. It is one of the important guidelines of China's economic and social development and is also the action guide that scientific and technological work must conform to in new stage and new phase. In the guidance of the concept of scientific development, we should strengthen the deployment of strategic areas essential to China's future development, allocate scientific and technological innovative resources rationally, build 2 to 3 more comprehensive experimental demonstration areas (cities) with multiple objectives optimization (City), form the layout of science and technology to support China's sustainable development, from directions of disciplines, research strength regional distribution and basic research, high technology innovation and their applications, and make new breakthroughs in "support" and "guide".

10.4 Establish National Ecological and Environmental Three-Dimensional Monitoring – Data Integration – System Simulation Trinity Experimental Center

In the world, more and more large-scale, long time series of environmental, socio-economic and health data can be obtained. These data have complex systemic characteristics and are difficult to be predicted. Only by elements of the system, studies are difficult to reveal the overall behavior of complex systems. There is a need to set up a three-dimensional national ecological and environmental monitoring – data integration – the system simulation of trinity experimental center, develop and deploy distributed intelligent sensor network system, develop a series of numerical models used in scientific research, develop effective mathematics and statistical tools with strong applicability, explore new methods applied in complex adaptive systems research, and enhance comprehensive integration capabilities of different models and data sets.

10.5 Cooperate with Neighboring Countries to Establish Global Change Asia Center

The integrated global research idea has promoted the increasing internationalization of ecological and environmental research. We should actively participate in international research projects, focus on organizing and carrying out such international cooperation research that takes China as its main body and long term scientific project as its center, closely related to China's needs, acquire relevant research data, and make the study area based on China and extent to the world. We will unite neighboring countries to establish Global Change Asia Center, a joint laboratory, and field observation stations, expand study areas, and provide decision support for safeguarding China's territorial security and environmental security.

References

[1] Ministry of Environmental Protection of the People's Republic of China. National 'Eleventh Five-Year' Science and Technology Development Plan for Environmental Protection. 2006.

[2] National Science Library, Chinese Academy of Sciences. Scientific research, dynamic monitoring Letters- Resources and Environment Science Album. The United Kingdom National Environment Research Council will launch a new plan of action for a subject. 2008, (92): 11–12.

[3] National Science Library, Chinese Academy of Sciences. Scientific research, dynamic monitoring Letters- Resources and Environment Science Album. South Korea released the 2nd comprehensive plan for environmental technology development. 2008, (92): 7–9.

[4] National Science Library, Chinese Academy of Sciences. Scientific research, dynamic monitoring Letters- Resources and Environment Science Album. The USA Long Term Ecological Research Program: Reconstruction of Network Science. 2008, (95): 7–9.

[5] National Science Library, Chinese Academy of Sciences. Scientific research, dynamic monitoring Letters- Resources and Environment Science Album. The USA Long Term Ecological Research Program Strategy Planning. 2008, (72): 1–7.

[6] National Science Library, Chinese Academy of Sciences. Scientific research, dynamic monitoring Letters- Resources and Environment Science Album. The USA National Ecological Observatory Network's China strategy. 2008, (95): 9–12.

[7] National Science Library, Chinese Academy of Sciences. Scientific research, dynamic monitoring Letters- Resources and Environment Science Album. The European Commission issued its Environmental Technology Action Plan 2005–2006 Annual Report. 2007, (66): 1–6.

[8] National Science Library, Chinese Academy of Sciences. Scientific research, dynamic monitoring Letters- Resources and Environment Science Album. United Kingdom Natural Environment Research Council 2005–2006 Strategic Plan. 2007, (62): 7–9.

[9] National Science Library, Chinese Academy of Sciences. Scientific research, dynamic monitoring Letters- Resources and Environment Science Album. The Research Theme of the United Kingdom Natural Environment Research Council 2007–2012 Strategy Plan. 2008, (78): 1–8.

[10] Resources and Environment. Chinese Academy of Science and Technology Agency. 2006.

[11] The People's Republic of China State Council. Outlined plan for China's Mid- to Long-term Science and Technology Development (2006–2020). 2006.

[12] The Ministry of Science and Technology of the People's Republic of China. National "Eleventh Five-Year" Science and Technology Development Plan. 2006.

[13] Z. G. Jang. Criteria for Classification of Endangered Species and the Preservation of Species. Bulletin of Biology.2000, 35 (8): 1–5.

SPRINGER NATURE

GPSR Compliance

The European Union's (EU) General Product Safety Regulation (GPSR) is a set of rules that requires consumer products to be safe and our obligations to ensure this.

If you have any concerns about our products, you can contact us on ProductSafety@springernature.com

In case Publisher is established outside the EU, the EU authorized representative is:

Springer Nature Customer Service Center GmbH
Europaplatz 3
69115 Heidelberg, Germany

MIX
Papier | Fördert gute Waldnutzung
FSC® C083411

Kolibri 360 GmbH
Ferdinand-Jühlke-Straße 7
99095 Erfurt, Deutschland
produktsicherheit@kolibri360.de